AF531553

PHYSIOLOGY OF PATHOGEN

RECENT ADVANCES IN PLANT PATHOLOGY SERIES

PHYSIOLOGY OF PATHOGEN

K. M. CHANDNIWALA

ANMOL PUBLICATIONS PVT. LTD.

NEW DELHI-110 002 (INDIA)

ANMOL PUBLICATIONS PVT. LTD.

Regd. Office: 4360/4, Ansari Road, Daryaganj,
New Delhi-110002 (India)
Tel.: 23278000, 23261597, 23286875, 23255577
Fax: 91-11-23280289
Email: anmolpub@gmail.com
Visit us at: www.anmolpublications.com

Branch Office: No. 1015, Ist Main Road, BSK IIIrd Stage
IIIrd Phase, IIIrd Block, Bangalore-560 085 (India)
Tel.: 080-41723429 • Fax: 080-26723604
Email: anmolpublicationsbangalore@gmail.com

Physiology of Pathogen

First Edition, 1995

Reprint, 1998, 2010

ISBN 81-7488-273-1

PRINTED IN INDIA

Printed at Mehra Offset Press, Delhi.

Preface

The present title has been intended primarily for the use of students enrolled in advanced undergraduate and postgraduate courses in plant pathology. The need of a textbook in this field was a pressing one since not only did the students find literature on the diseases and other pathological conditions difficult to obtain and to access, but no reasonably complete treatment of the subject from the student's stand point was to be had. To fill this need is the first objective of this title.

Considering the field as a whole, however, it was realised that in addition to a textbook for classroom use, the subject was also greatly in need of a reference work. The pathologists and the research worker in biology have not had an available guide to all phases of the subject of plant pathology as such for most of the information the research worker has been faced to consult are widely scattered in often almost inaccessible articles or incomplete accounts in larger volumes. Accordingly, the authors, turned their attention to the preparation of the present book, which deals specially with plant pathology, i.e. with the microbial diseases as well as with certain amicrobial deseases, injuries and abnormalities. The task has not been easy one.

Attempts have been made to present uptodate but brief introduction to principles underlying various stages in the development of plant diseases with special emphasis on their control. The language of the book has been painfully kept as simple and lucid style as possible. The subject matter has been interwoven for keeping up interest and curiosity of readers.

There can be no claim to originality except in the manner of treatment and much of the information has been obtained from the books and scientific journals available in different libraries.

Constructive criticism and suggestion for improvement of the book will be thankfully acknowledged.

Author

Preface

The present title has been intended primarily for the use of students enrolled in advanced undergraduate and postgraduate courses in plant pathology. The need of a textbook in this field was a pressing one since not only did the students find literature on the diseases and other pathological conditions difficult to obtain and to access, but no reasonably complete treatment of the subject from the student's stand point was to be found. To fill this need is the first objective of this title.

Considering the field as a whole, however, it was realised that in addition to a textbook for classroom use, the subject was also greatly in need of a reference work. The pathologists and the research worker in biology have not had an available guide to all phases of the subject of plant pathology as such. For most of the information the research worker has been faced to consult an widely scattered and often almost inaccessible articles or papers, the accounts in larger references. Accordingly, the authors turned their attention to the preparation of the present book, which deals specially with plant pathology. It treats the important diseases as well as with certain abnormal diseases, injuries and abnormalities. The task has not been easy one.

Attempts have been made to present uptodate but brief introductions to principles underlying various steps in the development of plant diseases with special emphasis on their control. The language of the book has been carefully kept as simple and lucid as possible. The subject matter has been thoroughly gone keeping up interest and continuity of matters.

There can be no claim to originality except in the manner of treatment and much of the information has been obtained from the books and scientific journals available in different libraries.

Constructive criticism and suggestion for improvement of the book will be thankfully acknowledged.

Author

Contents

1

EFFECTS OF PATHOGENS ON PLANT PHYSIOLOGICAL FUNCTIONS

EFFECTS OF PATHOGENS ON PHOTOSYNTHESIS

Photosynthesis is the basic function of green plants that enables them to transform light energy into chemical energy, which they can then utilize in their cell activities. Photosynthesis is the ultimate source of all energy used in plant or animal cells, since in a living cell all activities except photosynthesis expend the energy provided by photosynthesis.

In photosynthesis, carbon dioxide from the atmosphere and water from the soil are brought together in the chloroplasts of the green parts of plants and, in the presence of light, react to form glucose with concurrent release of oxygen:

$$6\,CO_2 + 6H_2O \xrightarrow[\text{chlorophyll}]{\text{light}} C_6H_{12}O_6 + 6\,O_2$$

In view of the fundamental position of photosynthesis in the life of plants, it is apparent that any interference by pathogens with photosynthesis results in a diseased condition in the plant. That pathogens do interfere with photosynthesis is obvious from the chlorosis they cause on many infected plants, from the necrotic lesions or large necrotic areas they produce on green plant parts, and from the reduced growth and amounts of fruits produced by many infected plants.

In leaf spot, blight, and other kinds of diseases in which there is destruction of leaf tissue or defoliation, photosynthesis is reduced because the photosynthetic surface of the plant is lessened. Even in other diseases, however, plant pathogens reduce photosynthesis especially in the late stages of diseases, by affecting the chloroplasts and causing their degeneration. The overall chlorophyll content of leaves in many fungal and bacterial diseases is reduced, but the photosynthetic activity of the remaining chlorophyll seems to remain unaffected. In some fungal and bacterial diseases, photosynthesis is reduced because the toxins, such as tentoxin and tabtoxin, produced by these pathogens inhibit some of the enzymes that are directly or indirectly involved in photosynthesis. In plants infected by many vascular pathogens, stomata remain partially closed, chlorophyll is reduced, and photosynthesis stops even before the plant eventually wilts. Most virus, mycoplasma, and nematode diseases induce varying degrees of chlorosis. In the majority of such diseases, photosynthesis of infected plants is reduced greatly. In advanced stages of disease, the rate of photosynthesis is no more than one-fourth the normal rate.

EFFECT OF PATHOGENS ON TRANSLOCATION OF WATER AND NUTRIENTS IN THE HOST PLANT

All living plant cells require an abundance of water and an adequate amount of organic and inorganic nutrients in order to live and to carry out their physiological functions. Plants absorb water and inorganic (mineral) nutrients from the soil through their root system. These substances are generally translocated upward through the xylem vessels of the stem and into the vascular bundles of the petioles and leaf veins, from which they enter the leaf cells. The minerals and part of the water are utilized by the leaf and other cells for synthesis of the various plant substances, but most of the water evaporates out of the leaf cells into the intercellular spaces and from these diffuses into the atmosphere through the stomata. On the other hand, nearly all organic nutrients of plants are produced in the leaf cells, following photosynthesis, and are translocated downward and distributed to all the living plant cells by passing for the most part through the phloem tissues. When a pathogen interferes with the upward movement of inorganic nutrients and water or with the downward movement of organic substances, diseased conditions will result in the parts of the plant denied these materials. The diseased parts, in turn, will be unable to carry out their own functions and will deny the rest of the plant their services or their products, thus causing disease of the entire plant. For example, if water movement to the leaves is inhibited, the leaves cannot function properly, photosynthesis is reduced or stopped, and few or no nutrients are available to move to the roots, which in turn become starved, diseased, and may die.

Interference with Upward Translocation of Water and Inorganic Nutrients

Many plant pathogens interfere in one or more ways with the translocation of water and inorganic nutrients through plants. Some pathogens affect the integrity or function of the roots, causing them to absorb less water, other pathogens, by growing in the xylem vessels or by other means, interfere with the translocation of water through the stem; and, in some diseases, pathogens interfere with the water economy of the plant by causing excessive transpiration through their effects on leaves and stomata.

Effect on Absorption of Water by Roots. Many pathogens, such as the damping-off fungi, the root-rotting fungi and bacteria, most nematodes, and some viruses cause an extensive destruction of the roots before any symptoms appear on the above ground parts of the plant. Root injury directly affects the amount of functioning roots and decreases proportionately the amount of water absorbed by the roots. Some vascular parasites, along with their other effects, seem to inhibit root hair production, which reduces water absorption. These and other pathogens also alter the permeability of root cells, an effect that further interferes with the normal absorption of water by roots.

Effect on Translocation of Water through the Xylem. Fungal and bacterial pathogens that cause damping-off, stem rots, and cankers may reach the xylem vessels in the area of the infection and, if the affected plants are young, may cause their destruction and collapse. Affected vessels may also be filled with the bodies of the pathogen and with substances secreted by the pathogen or by the host in response to the pathogen, and may become clogged. Whether destroyed or clogged, the affected vessels cease to function properly and allow little or no water to pass through them. Certain pathogens, such as the crown gall bacterium (*Agrobacterium tumefaciens*), the clubroot fungus (*Plasmodiophora brassicae*), and the root-knot nematode (*Meloidogyne* sp.)

induce gall formation in the stem or roots or both. The enlarged and proliferating cells near or around the xylem exert pressure on the xylem vessels, which may be crushed and dislocated, thereby becoming less efficient in transporting water.

The most typical and complete dysfunction of xylem in translocating water, however, is observed in the vascular wilts caused by the fungi *Ceratocystis, Fusarium*, and *Verticillium*, and bacteria like *Pseudomonas* and *Erwinia*. These pathogens invade the xylem of roots and stems and produce diseases primarily by interfering with the upward movement of water through the xylem. In many plants infected by these pathogens the water flow through the stem xylem is reduced to a mere 2 to 4 percent of that flowing through the stems of healthy plants. In general, the rate of flow through infected stems seems to be inversely proportional to the number of vessels blocked by the pathogen and by the substances resulting from the infection. Evidently, more than one factor is usually responsible for vascular dysfunction in the wilt diseases. Although the pathogen is the single cause of the disease, some of the factors responsible for the disease syndrome originate directly from the pathogen, while others originate from the host in response to the pathogen. The pathogen can reduce the flow of water through its physical presence in the xylem as mycelium, spores, or bacterial cells and by production of large molecules (polysaccharides) in the vessels. The infected host may reduce the flow of water through reduction in the size or collapse of vessels due to infection, development of tyloses in the vessels, release of large-molecule compounds in the vessels as a result of cell wall breakdown by pathogenic enzymes, and reduced water tension in the vessels due to pathogen-induced alterations in foliar transpiration.

Effect on Transpiration. In plant diseases in which the pathogen infects the leaves, transpiration is usually increased. This is the result of destruction of at least part of the protection afforded the leaf by the cuticle, an increase in permeability of leaf cells, and the dysfunction of stomata. Disease like the rusts, mildews, and apple scab destroy a considerable portion of the cuticle and epidermis, and this results in uncontrolled loss of water from the affected areas. If water absorption and translocation cannot keep up with the excessive loss of water, loss of turgor and wilting of leaves follows. The suction force of excessively transpiring leaves is abnormally increased and may lead to collapse or dysfunction of underlying vessels through production of tyloses and gums.

Interference with the Translocation of Organic Nutrients through the Phloem

Organic nutrients produced in leaf cells through photosynthesis move through plasmodesmata into adjoining phloem elements. From there they move down the phloem sieve tubes and eventually, again through plasmodesmata, into the protoplasm of living nonphotosynthetic cells, where they are utilized, or into storage organs, where they are stored. Thus, in both cases, they are removed from "circulation." Plant pathogens may interfere with the movement of organic nutrients from the leaf cells to the phloem or with their translocation through the phloem elements and, possibly, with their movement from the phloem into the cells that will utilize them.

Obligate fungal parasites, such as the rust and mildew fungi, cause an accumulation of photosynthesis products, as well as inorganic nutrients, in the areas invaded by the pathogen. In these diseases, the infected areas are characterized by reduced photosynthesis and increased respiration. However, synthesis of starch and of other compounds as well as dry weight are temporarily increased in the infected areas, indicating translocation of organic nutrients from uninfected areas of the leaves or from healthy leaves toward the infected areas.

In some virus diseases, particularly the leaf-curling type and some yellows diseases, starch accumulation in the leaves is a common phenomenon. In most of these diseases, starch accumulation in the leaves is mainly the result of degeneration (necrosis) of the phloem of infected plants, which is one of the first symptoms. It is also possible, however, at least in some virus diseases, that the interference with translocation of starch stems from inhibition by the virus of the enzymes that break down starch into smaller translocatable molecules. This is suggested by the observation that in some mosaic diseases, in which there is no phloem necrosis, infected, discolored areas of leaves contain less starch than "healthy", greener areas at the end of the day, a period favorable for photosynthesis; but the same leaf areas contain more starch than the "healthy" areas after a period in the dark, which favors starch hydrolysis and translocation. This suggests not only that virus-infected areas synthesize less than healthy ones, but also that starch is not easily degraded and translocated from virus-infected areas, although no damage to the phloem is present.

EFFECT OF PATHOGENS ON HOST PLANT RESPIRATION

Respiration is the process by which cells, through enzymatically controlled oxidation (burning) of the energy-rich carbohydrates and fatty acids, liberate energy in a form that can be utilized for the performance of various cellular processes. Plant cells carry out respiration in, basically, two steps. The first step involves the degradation of glucose to pyruvate and is carried out, either in the presence or in the absence of oxygen, by enzymes found in the ground cytoplasm of the cells. The production of pyruvate from glucose follows either the *glycolytic pathway*, otherwise known as *glycolysis*, or to a lesser extent, the *pentose pathway*. The second step involves the degradation of pyruvate, however produced, to CO_2 and water. This is accomplished by a series of reactions known as the *Krebs cycle*, which is accompanied by the so-called *terminal oxidation* and is carried out in the mitochondria only in the presence of oxygen. Under normal (aerobic) conditions, that is, in the presence of oxygen, both steps are carried out, and one molecule of glucose yields, as final products, six molecules of CO_2 and six molecules of water,

$$C_6H_{12}O_6 + 6O_2 \rightarrow 6CO_2 + 6H_2O$$

with concomitant release of energy (678,000 calories). Some of this energy is lost, but almost half is converted to 20–30 reusable high-energy bonds of adenosine triphosphate (ATP). The first step of respiration contributes two ATP molecules per mole of glucose, and second step contributes the rest. Under anaerobic conditions, however—that is, in the absence of oxygen—pyruvate cannot be oxidized, but it instead undergoes *fermentation* and yields lactic acid or alcohol. Since the main process of energy generation is cut off, for the cell to secure the necessary energy a much greater rate of glucose utilization by glycolysis is required in the absence of oxygen than is in its presence.

The energy-storing bonds of ATP are formed by the attachment of a phosphate (PO_4) group to adenosine diphosphate (ADP), at the expense of energy released from the oxidation of sugars. The coupling of oxidation of glucose with the addition of phosphate to ADP to produce ATP is called *oxidative phosphorylation*. Any cell activity that requires energy utilizes the energy stored in ATP by simultaneously breaking down ATP to ADP and inorganic phosphate. The presence of ADP and phosphate in the cell, in turn, stimulates the rate of respiration. If, on the other hand, ATP is not utilized sufficiently by the cell for some reason, there is little or no regeneration of ADP and respiration is slowed down. The amount of ADP (and phosphate) in the cell is determined, therefore, by the rate of energy utilization; this rate, in turn, determines the rate of respiration in plant tissues.

The energy produced through respiration is utilized by the plant for all types of cellular work, such as accumulation and mobilization of compounds, synthesis of proteins, activation of enzymes, cell growth and division, defense reactions, and a host of other processes. The complexity of respiration, the number of enzymes involved in respiration its occurrence in every single cell, and its far-reaching effects on the functions and existence of the cell make it easy to understand why respiration of plant tissues is one of the first functions to be affected when plants are infected by pathogens.

Respiration of Diseased Plants

When plants are infected by pathogens, the rate of respiration generally increases. This means that affected tissues use up their reserve carbohydrates faster than healthy tissues would. The increased rate of respiration appears shortly after infection—certainly by the time of appearance of visible symptoms—and continues to rise during the multiplication and sporulation of the pathogen. After that, respiration declines to normal levels or to levels even lower than those of healthy plants. Respiration increases more rapidly in infections of resistant varieties, in which large amounts of energy are needed and used for rapid production or mobilization of the defense mechanisms of the cells. In resistant varieties, however, respiration also declines quickly after it reaches its maximum. In susceptible varieties, in which no defense mechanisms can be mobilized quickly against a particular pathogen, respiration increases slowly after inoculation, but it continues to rise and remains at a high level for much longer periods.

Several changes in the metabolism of the diseased plant accompany the increase in respiration after infection. Thus, the activity or concentration of several enzymes of the respiratory pathways seem to be increased. The accumulation and oxidation of phenolic compounds, many of which are associated with defense mechanisms in plants, are also greater during increased respiration. Increased respiration in diseased plants is also accompanied by an increased activation of the pentose pathway, which is the main source of phenolic compounds. Increased respiration is sometimes accompanied by considerably more fermentation than observed in healthy plants, probably as a result of an accelerated need for energy in the diseased plant under conditions in which normal aerobic respiration cannot provide sufficient energy.

The increased respiration in diseased plants is apparently brought about, at least in part, by uncoupling of the oxidative phosphorylation. In that case, no utilizable energy (ATP) is produced through normal respiration in spite of the use of existing ATP and accumulation of ADP, which stimulates respiration. The energy required by the cell for its vital processes is then produced through other, less efficient ways including the pentose pathway and fermentation.

The increased respiration of diseased plants can also be explained as the result of increased metabolism. In many plant diseases, growth is at first stimulated, protoplasmic streaming increases, and materials are synthesized, translocated, and accumulated in the diseased area. The energy required for these activities derives from ATP produced through respiration. The more ATP is utilized, the more ADP is produced and further stimulates respiration. It is also possible that the plant, because of the infection, utilizes ATP energy less efficiently than a healthy plant. Because of the waste of part of the energy, an increase in respiration is induced and the resulting greater amount of energy enables the plant cells to utilize sufficient energy to carry out their accelerated processes.

Although oxidation of glucose via the glycolytic pathway is by far the most common way through which plant cells obtain their energy, part of the energy is produced via the pentose pathway. The latter seems to be an alternate pathway of energy production to which plants resort under conditions of stress. Thus, the pentose pathway tends to replace the glycolytic pathway as the plants grow older and differentiate, and to increase upon treatment of the plants with hormones, toxins, wounding, starvation, and so on. Infection of plants with pathogens also tends, in general, to activate the pentose pathway over the level at which it operates in the healthy plant. Since the pentose pathway is not directly linked to ATP production, the increased respiration through this pathway fails to produce as much utilizable energy as the glycolytic pathway and is, therefore, a less efficient source of energy for the functions of the diseased plant. On the other hand, the pentose pathway is the main source of phenolic compounds, which play important roles in the defense mechanisms of the plant against infection.

Effect of Pathogens on Permeability of Cell Membranes

Cell membranes consist of a double layer of lipid molecules in which are embedded many kinds of protein molecules, parts of which usually protrude on both sides of the lipid bilayer. Membranes function as permeability barriers that allow passage into a cell only of substances the cell needs and inhibit passage out of the cell of substances needed by the cell. The lipid bilayer is impermeable to most biological molecules. Small water-soluble molecules such as ions (charged atoms or electrolytes), sugars, and amino acids flow through or are pumped through special membrane channels made of proteins. In plant cells, because of the cell wall, only small molecules reach the cell membrane. In animal cells, however, and in artificially prepared plant protoplasts, large molecules or particles may also reach the cell membrane and enter the cell by *endocytosis:* A patch of the membrane surrounds and forms a vesicle around the material to be taken in, brings it in and released it inside the cell. Disruption or disturbance of the cell membrane by chemical or physical factors alters (usually increases) the permeability of the membrane with subsequent uncontrollable loss of useful substances as well as inability to inhibit the inflow of undesirable substances or excessive amounts of any substances.

Changes in cell membrane permeability are often the first detectable responses of cells to infection by pathogens, to most host-specific and several nonspecific toxins, to certain pathogen enzymes, and to certain toxic chemicals, such as air pollutants. The most commonly observed effect of changes in cell membrane permeability is the loss of *electrolytes,* that is, of small water-soluble ions and molecules from the cell. It is not certain, however, whether the cell membrane is the initial target of pathogen toxins and enzymes and that the accompanying loss of electrolytes is the initial effect of changes in cell membrane permeability, or whether the pathogen products actually affect other organelles or reactions in the cell, in which case cell permeability changes and loss of electrolytes are secondary effects of the initial events. If pathogens do, indeed, affect cell membrane permeability directly, it is likely that they bring this about by stimulating certain membrane-bound enzymes, such as ATPase, which are involved in the pumping of H^+ in and K^+ out through the cell membrane, by interfering with processes required for maintenance and repair of the fluid film making up the membrane, or by degrading the lipid or protein components of the membrane by pathogen-produced enzymes.

EFFECTS OF PATHOGENS ON TRANSCRIPTION AND TRANSLATION

Transcription of cellular DNA into messenger RNA and translation of messenger RNA to produce proteins are two of the most basic, general, and precisely controlled processes in the biology of any normal cell (Fig. 1.1). The part(s) of the genome involved and the level and timing of transcription and translation vary with the stage of development and the requirements of each cell. Nevertheless, disturbance of any one of these processes, by pathogens or environmental factors, may by its effect on the expression of genes cause drastic unfavourable changes in the structure and function of the affected cells.

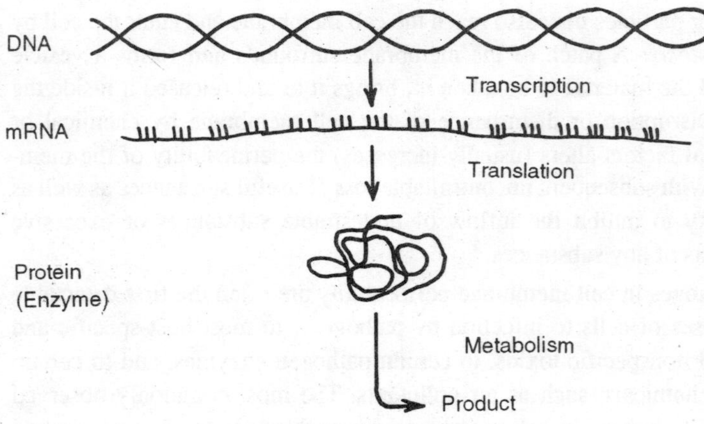

Fig. 1.1. Transcription and translation processes.

Effect on Transcription

Several pathogens, particularly viruses and fungal obligate parasites such as the rusts and powdery mildews, affect the transcription process in infected cells. In some cases, pathogens affect transcription by changing the composition, structure, or function of the chromatin associated with the cell DNA. In some diseases, especially those caused by viruses, the pathogen, through its own enzyme or by modifying the host enzyme (RNA polymerase) that makes RNA, utilizes the host cell nucleotides and machinery to make its own (rather than host) RNA. In several diseases, the activity of ribonucleases—enzymes that break down RNA—is increased, perhaps by formation in infected plants of new kinds of ribonucleases not known to be produced in healthy plants. Finally, in several diseases, infected plants, particularly resistant ones seem to contain higher levels of RNA than healthy plants, especially in the early stages of infection. It is generally believed that greater RNA levels and, therefore, increased transcription in cells indicate increased synthesis of substances involved in the defense mechanisms of plant cells.

Effect on Translation

Infected plant tissues often have increased activity in several enzymes, particularly those associated with generation of energy (respiration) or with production or oxidation of various phenolic compounds, some of which may be involved in (defense) reactions to infection. Although a certain amount of some of these enzymes (proteins) may be present in the cell at the time of infection, several are produced *de novo*, necessitating increased levels of transcription and of RNA polymerase activity. As mentioned above, ribonuclease synthesis and activity also increase in infected tissues. Increases in protein synthesis in infected tissues have been observed primarily in hosts resistant to the pathogen and reach their highest levels in the early stages of infection, that is, in the first 2–20 hours after inoculation. If resistant tissues are treated before or during infection with inhibitors of protein synthesis, their resistance to the pathogen is reduced. These observations suggest that much of the increased protein synthesis in plants attacked by pathogens reflects the increased production of enzymes and other proteins that is involved in the defense reactions of plants.

2

EFFECT ON TRANSPIRATION AND WATER BALANCE

Life, in all probability, originated in water and in its primary state is adapted to function only in an aquatic medium, at least, at its saturation level, if not immersed in it. As the various organisms, both plants and animals, extended their habitat to life on land, they were continually subjected to the threat of desiccation and death by forces of nature, against which they retain water in themselves, but the struggle to counterbalance this threat seems to be, in a large measure, the essence of life. The title of this chapter indicates only this aspect of the life process and the problem assumes a greater significance when we are dealing with organisms put to an additional disadvantage—the "disease" which is attended by various repercussions on key metabolic functions.

Among the most important constituents of plants, water is quantitatively by far the most abundant. Protoplasm, the physical basis of life, is itself a hydrated gel, containing water, no less than 80–90% of its total weight in its active stage. It follows, therefore, that any disturbance to the normal process of life would have a marked effect on this component. The importance of water can be better appreciated if we list its various roles in the life processes of plants. In addition to its being an essential constituent of living protoplasm, it forms a continuous phase permeating the entire plant body and it takes part in various chemical reactions, particularly hydrolysis and photosynthesis. It also plays an essential role in maintaining the form and structure of herbaceous plant tissues through the maintenance of cell turgidity. The various physiological functions of protoplasm itself are controlled by the extent of its hydration. It would, thus, seem clear that any disturbance to the water economy would naturally lead to various physical and physiological consequences.

The bulk of water in plants is not in a static condition but is part of a hydrodynamic system which operates as one unit and is composed of the balancing forces at the cellular level. The water content of the various cells and tissues is in a continuous state of change. To understand fully how and why such changes in water content occur, requires consideration of the chemical and physical characteristics of various parts of the cell, of the forces which hold water in cells and of the principles which govern water movement between cells and their environment.

Water is held in the cells of plants, principally by osmotic and imbibitional forces. Osmotic forces are developed by the presence of solutes which decrease the activity or free energy of water molecules, resulting in a decrease in its diffusion pressure. Imbibition occurs because of the attraction for water of various hydrophilic colloids such as cellulose and proteins. In this case, water is held in the microcapillary lattices of complex molecules and held largely by surface forces. Thus, the free energy of water is reduced severally in plant cells and the amount by which the diffusion pressure of water is reduced is termed the diffusion pressure deficit (DPD). The DPD of a solution is equal to its osmotic pressure, but in plant cells this is greatly modified by the wall pressure (which is equal and opposite to turgor pressure) and may decrease from a value almost equal to the osmotic pressure of the cell sap (when turgor pressure is equal to zero) to zero as the turgor and wall pressures increase. The DPD of a cell would, therefore, be equal to the osmotic pressure of a solution in which it neither gains nor loses water. In other words, the water content of the cell remains unchanged as long as the DPD of results in a movement of water from within or without the cell. Besides these osmotic forces, there is an "active" absorption of water by the cells which involves expenditure of energy and this resembles the accumulation of ions. Loss of water from the cells would, thus, involve a disruption of the equilibrium that contributes to its retention in them.

Disease manifests itself primarily in various functional disturbances, most of which, if not all, can be traced either directly or indirectly to a disturbed water balance in plants. This fact becomes evident when we study the various derangements attending on disease and trace their bearing on the water economyy of the individual. But, to appreciate this, it is necessary to have a basic picture of the various means by which a plant strives to keep up its water balance and maintains its norm. Excellent reviews and treatises have appeared on this subject.

I. WATER BALANCE IN NORMAL HEALTHY PLANTS

The maintenance of a favorable water balance in plants demands that the loss of water from the leaves by transpiration (guttation loss being negligible in most cases) shall not, except for very short periods, exceed the supply of water to them. If it does, the total volume of water in plants is reduced and a thirst develops, the most obvious symptom being wilting. If, therefore, thirst and its consequences, as we shall discuss later, are to be avoided a regular flow of water to the leaves is essential. Normally, plants obtain water from the soil through their root system. In its passage from the root hairs to the foliage, water encounters resistance. An adequate supply depends on overcoming this resistance, which varies in magnitude from plant to plant and, probably, from time to time within the same plant.

Absorption of water is influenced by conditions that affect the metabolism and permeability of root cells, such as low oxygen tension, high CO_2 concentration and also the total soil moisture stress at the root surface. It is also accentuated by the factors of root density and root elongation which determine the distance water has to move and so influence the total amount of water available. The rate of transpiration determines the extent of water deficit and, hence, the amount of water needed to attain equilibrium. In most plants studied, there is a distinct lag in the rate of absorption of water as compared with the rate of transpiration during daylight hours, showing a condition of internal water deficit; thus seems to be of almost daily occurrence in most plants during their growing season and the mechanism by which plants attain reduction of internal water deficit constitutes the dynamics of its water relations. Naturally, the forces directed to achieve this end should tend to decrease transpirational water loss and increase absorption of water from the substratum, conduction being usually incidental to these two factors. Conversely, forces that tend to disturb the normal water balance act on either of these two, or both, and thereby increase the internal water deficit, causing thirst.

It seems to be generally accepted that the mechanism of transpiration involves evaporation of water either through the cuticle or through the stomates. Consequent on this evaporation, a gradient is set up between the evaporating surface and the trachea and water moves from the latter to the former. This process continues only as long as the gradient is maintained, which is subject to the influence of the intervening living cells. The DPD of the evaporating surface is regulated to a large extent by the opening and closing of stomata. That stomata respond to stimuli, such as light and darkness, changes in leaf water content, temperature, CO_2 content of the air, shock stimuli, changes in H-ion concentration and ionic effects, seems well established. The greater the DPD of the tracheal contents, the less the readiness with which water moves to the evaporating surface and the more the withdrawal of water into the intermicellar spaces of the cell wall at the evaporating surface, thus greatly increasing resistance to evaporation.

Transpiration proceeds as long as the suction force at the evaporating surface exceeds that of the leaf cells and the trachea. If the leaf cells are to remain fully turgid, their suction force must be greater than that in the tracheae on the one hand, and that at the evaporating surface, on the other. Increase in either of the latter two leads to loss of turgor in leaf cells, and wilting follows. Normally, the osmotic pressure of leaf cells is kept up at a high level by the reduction of water by transpiration, on the one hand, and by accumulation of photosynthates, on the other. Any decrease in water in leaf cells produces changes, such as hydrolysis of starch and protein, which give rise to osmotically active substances resulting in a flow of water from the trachea, creating a tension in the xylem. As long as the DPD of soil water and that of leaf cells exceeds both, absorption is facilitated. The significance of a high DPD of leaf cells is at least twofold. It causes the development of a high tension in the xylem, which is necessary for absorption against the resistance of root cells; at the same time it reduces transpiration. Thus, absorption of water tends to decrease the DPD gradient between the leaf and the tracheal fluid, whereas transpiration increases it. The tensions disappear during periods of good water supply and low transpiration; hence, existence of tension can be taken to indicate some internal water deficit, which, however, is prevented from becoming serious by the action of tension itself on absorption and transpiration.

Thus, we have here a self-regulatory mechanism for preventing the development of a serious internal water deficit. The efficiency of this mechanism is limited by the osmoregulatory changes of leaf cells. It is precisely in those plants whose cell sap has a high osmotic pressure that there is possibility of considerable tension developing when conditions favor high transpiration. This in itself is likely to facilitate the intake of water by the roots and postpone development of a serious water deficit in the leaves. It would, thus, be evident that the daily variations in osmotic pressure of plant cells, particularly in the leaf, are most significant in that they swing back the water status of the plant to a condition of balance from that of imbalance. The cycle of variations in the osmotic quantities may be quite different under environmental conditions departing from the normal, which favors high transpiration in the presence of an adequate water supply. Seasonal variations are also known in the case of perennial plants.

II. GENERAL SYMPTOMS AND NATURE OF WATER DEFICIENCY

The most apparent effect of water deficit is loss of form in herbaceous plant tissues—as a sequel to loss of turgor—such as wilting, flagging, and drooping of leaves and stem tips. Root hairs are also believed to wilt very commonly, although such wilting cannot be ordinarily observed. With xerophytes and plants with rigid leaves, wilting may not be apparent, although physiologically comparable conditions exist in them. Wilting occurs when the turgor pressure is almost zero and represents a stage when the DPD is equal to the average osmotic pressure. A temporary excess of transpiration over absorption results in such changes, but normally plants regain their form as more water is made available.

Of all the major plant processes, none is more obviously affected by a deficiency of water than growth, although meristematic tissues and young cells, because of their high imbibitional forces, are able to obtain water from older tissues. Water deficit interferes with cell enlargement and cell division, but promotes differentiation. Continued water deficit eventually results in cessation of growth. A decrease in normal water supply is reflected on growth largely because of a corresponding decrease in the production of auxins. This effect and other metabolic changes, detailed below, produce plants with a stunted appearance. Rosetting of leaves takes place from shortening of internodes. Leaf area is also greatly reduced. Reduction in root growth means decrease in available surface freely permeable to water. This is accentuated by the destruction of root hairs and premature suberization. In general, a reduced water content of a plant, whether caused by a shortage of water in the soil, or by high transpiration rates, results in a lower shoot: root ratio on a weight basis.

The effect of decrease in turgor on transpiration, unlike other vital functions, seems to be indirect. Transpiration is basically a passive process and as such is determined largely by the diffusion gradient from leaf to atmosphere and the rate of water supply to the roots. Leaf water content itself is not likely to affect transpiration unless severe wilting occurs; it depends on the sensitivity of the stomates to changes in turgor. A correlation between the rate of transpiration and the width of the stomatal opening is feasible only when the water content is sufficient, although stomatal behavior is, doubtless, influenced by other factors.

Reduction in leaf water content usually results in a diminished rate of photosynthesis and this is pronounced before any wilting occurs. A loss of water of 16 to 47% or more causes a decrease of 20% in the rate of photosynthesis, but there seems to be no correlation between the amount of water lost and the intensity of photosynthesis. In plants which have recovered from wilting, the ability to carry on carbon assimilation is not restored to normalcy, but is reduced by 35 to 59%. The influence of water deficit on photosynthesis may be direct, through a decrease in protoplasmic hydration, or indirect, through stomatal regulation. Stomates have an important role in the absorption of CO_2 from the air and in the evaporation of water. In general, water loss of even 10% induces stomatal closure. In certain species the sensitivity of the stomatal mechanism is so great, that a loss of even 3 to 5% results in their closure, e.g., *Vicia and Chrysanthemum.*

Decrease in water content of cells and tissues produces important changes in their physical and chemical properties. Dehydration produces changes in viscosity and permeability of protoplasm. Moderate dehydration increases viscosity and slackens Brownian movement. A slow adaptation (hardening) is exhibited when dehydration is gradual, whereas when the process of dehydration is drastic, complete gelatin occurs and the protoplasm becomes rigid and brittle, in which case the chances of recovery are remote. Northen (1943) and Stocker (1948) believed that dehydration causes dissociation of the protoplasm, resulting in the physical changes observed, followed by activation of certain enzymes and increased respiration. In general, hydrolytic processes are increased, e.g., hydrolysis of starch and proteins and synthetic processes are hindered, e.g., protein formation from amino acids.

A relatively high water content of the leaf tissues favors accumulation of starch at the expense of sugars in many species, while a reduction in water content favors the transformation of starch to sugars or polysaccharides. An increased supply of available nitrogen stimulates the utilization of carbohydrates, and if in addition sufficient moisture is available, growth and formation of new organs are accelerated. On the other hand, if sufficient moisture is not available, growth is interrupted and polysaccharides tend to accumulate. The breakdown of carbohydrates in leaves may be accompanied by their deposition in roots. This happens to be the case in a majority of plants. However, when wilting is slight, the changes detailed above are not noticed, until further desiccation stimulates these processes. It is, therefore, logical to presume that fluctuations of considerable magnitude in the water content of plants, due to their "inefficient" hydrodynamic system, result in changes which increase the proneness of the host to attacks by pathogenic, root-infecting organisms.

A. Physiological Wilting

In general, diminution of water content affects the leaf cells most, as compared with other parts of plants and this results in their partial or complete loss of turgor. Visible manifestations of wilting are frequent also in young succulent stem tips, floral parts, or even fruits and root hairs. The term "incipient wilting" is applied when the loss of turgor is not great enough to result in visible drooping. This condition is found in most terrestrial plants on bright and warm days. If the suction force of the evaporating surface is greater than that of the tracheal contents, and both are greater than that of the living cells, water may pass through the leaf cells without maintaining the latter in full turgidity. This seems to be borne out by the fact that even in a wilted leaf water continues to be transpired. The development of a considerable degree of tension in the tracheae would also cause a flow of water from the living leaf cells to the tracheae. Normally, tension in the xylem does not exist for a long time when there is sufficient water supply and time for recovery. If, however, this continues owing to a prolonged shortage of water supply, the leaves pass on to a state of permanent wilting through transient wilting. Both incipient wilting and transient wilting differ from permanent wilting in that the latter results not from a transitory excess of transpiration over water absorption, but from a deficiency of water in the soil. Plant do not recover from permanent wilting unless the water content of the soil in which they are rooted increases; they do not regain their turgor when placed in a saturated atmosphere. Permanent wilting marks a stage beyond which water is not available for normal plant functions. It usually arises when the soil water has fallen to such a low level that the plant cannot extract it. In other words, the total soil moisture stress increases.

This stress represents the DPD of soil water and is equivalent to the combined effect of the soil moisture tension and the osmotic concentration of the soil solution. As the soil water is depleted by absorption by the roots, an increase in the solute concentration at the root surface occurs. This is because the rate of solute absorption and the rate of water absorption are controlled by different factors. This causes a slackening in the rate of absorption, if the soil water is not replenished and wilting is hastened. In a soil slowly drying up, temporary wilting slowly grades over to permanent wilting. The nocturnal recovery of the plant from temporary wilting is achieved less and less completely until even the slightest recovery fails to take place. During permanent wilting the stress in the hydrodynamic system gradually becomes intensified; even if the stomates are closed, as they usually are in permanently wilted plants, cuticular transpiration continues, gradually reducing the total volume of water within the plant. Prolongation of this state for more than a few days results in the death of root hairs and, thereafter, recovery of the plant to the normal condition is slow. It also results in development of high tensions in the cells and these, in turn, subject protoplasm and cell walls to a centripetally directed pull which may lead to death of cells and ultimate wilting.

Physiological wilting of plants may also result from conditions other than soil moisture deficiency. Roots may be unable to absorb water from soil because of factors that interfere with absorption. Absorption of water by the roots is both active and passive. The former involves an expenditure of energy derived from respiration and resembles the phenomenon of salt accumulation. The latter is controlled by the osmotic gradient that exists between the sap in the xylem and soil water. Any interference with either, or both, results in a physiologic drought. These two processes are, however, interrelated and controlled by the metabolic state of root cells, since salt uptake causes osmotic absorption of water proposed that water intake and respiration are linked by transfer of energy through adenosine triphosphate (ATP), perhaps in a mechanism wehrein auxin molecules function as a part of the transport system. Although the nature of the relation between respiration and water intake is uncertain, inhibitors of the former have been found to reduce or prevent water absorption. Absorption is also impeded by flooding of the soil. This results in "flooding injury" which manifests itself in yellowing and wilting of leaves. These symptoms are attributed to desiccation caused by a decreased absorption. The picture is complicated further by the injury and death of root cells due to poor aeration. This, however, does not seem to explain wilting and death of shoots adequately, since plants can live for some days after their root systems are killed, if the soil is kept saturated with water. It, therefore, appears probable that the various symptoms produced by flooding have several causes, in addition to interference with absorption, such as inhibition of root growth, root development and elongation, translocation of toxic substances either released from dying cells (e.g., ethylene) or produced in the soil, and nonselective absorption of minerals as a sequel to the death of the root system. An imbalance of minerals is known to cause injury to leaves, such as mesophyll collapse noticed in citrus leaves.

B. Pathological Wilting

Pathological wilting, as the term denotes, is caused by pathogenic agencies. It occurs in diseases variously described as damping-off, die-back, foot rot, take-all, and wilt. Damping-off is caused by primitive parasites which attack the seedling of many plants and bring about their death by extensive rotting of the root and collar region (e.g., species of *Fusarium, Pythium,* and *Rhizoctonia*). Foot rot and take-all represent instances wherein the roots and crown regions are damaged and the plants exhibit characteristic symptoms of water deficiency before they die. In the case of foot rot of paddy (*Gibberella fujikuroi*), however, the presence of characteristic metabolites of the causal organism—fusaric acid and gibberellic acid, the former a wilt toxin and the latter having the property of growth substances—has been recently demonstrated. The final symptoms of the disease, in the case, would naturally arise not only from the interference with water uptake consequent to the damage to subterranean portions of the plant, but also as a result of the presence of these metabolites. Fusaric acid in low concentration increases permeability of protoplasts to water and reduces it at higher concentrations. Gibberellic acid, on the other hand, inhibits root growth among other things and this would aggravate root dysfunction. The syndrome of foot rot of paddy thus appears to be the product of an intricate interplay of diverse factors. In the case of take-all no such evidence has been brought forward so far and the disease appears to arise out of the dysfunction of root system resulting from injury by the causal organism, *Ophiobolus graminis*. Dieback of plants and branches is caused by definite interference with water supply, either by tissue disintegration or by inducing gum formation in the xylem region.

In vascular wilt diseases, however, there are to be seen both a local effect which leads to the necrosis of the directly affected tissues and a general systemic affect leading to the death of the entire plant. Wilting may occur in most instances, although it is not evident in such cases as cabbage yellows caused by *Fusarium conglutinans*. Most of the vascular wilts have many features in common, such as vascular discoloration, epinasty, yellowing, and vein clearing, and this suggests a common basis for their origin. There is a general derangement in the water balance observable in the infected plants, but the exact manner in which it is brought about still defies a precise answer.

Damage to the absorptive organs has sometimes been suggested as a principal factor in the disease, but this does not seem to be apparent. In most cases the extent of root damage is too meager to account for such an acute water shortage and often wilting occurs before any visible damage to the root is noticeable. Injury to roots causing death of root cells, will not interfere with the supply of water to the shoot, but a functional disturbance may lead to indiscriminate passage of toxic compounds and minerals through them and thereby cause manage to the shoot. Recent work on *Fusarium* wilt of cotton seems to provide evidence for such an occurrence.

There is considerable evidence that dysfunction of conductive elements causes an acute water shortage. This appears to be brought about by various physical and chemical causes, such as, the presence in the tubes of wefts of hyphae and masses of organisms (as in bacterial wilts) or gums, tyloses, gels, and gas pockets, all arising out of the chemical activities of the parasite during its interactions with the host. These may interfere with water flow through the xylem, either by obstructing the free passage or by increasing the viscosity of the tracheal fluid. However, in view of the present state of knowledge of the mechanism of sap flow through the xylem, these barriers seem to have only a limited significance. In addition to a general water shortage, there are symptoms such as veinclearing, epinasty, and necrosis in the leaves, which indicate the operation of a translocated systemic factor. The course of transpiration in diseased plants, and the exudations of minerals and amino acids from leaves, as shown by Linskens (1955), indicate that the leaf cells have lost their normal functions of osmoregulation. The role of pectic enzymes, initiating a chain of reactions leading to the release of phenols in the transpirational stream, is sometimes suggested to explain this derangement. Poisoning of leaf cells by metabolites of the causal agent carried along the sap stream is also suggested to account for this condition. The mechanism by which the toxin(s) brings about such changes appears to be primarily due to the destruction of the osmotic prereqistites for turgor, causing a release of cellular components into the outer medium, which appears to be the basis of pathological wilting in plants. This results in the loss of water-retaining capacity of the leaf cells and an efflux of water into the transpirational stream causing an increase in water loss. Mere shortage of water will not produce such effects as claimed by many; on the other hand, shortage of water may reduce the leaf area, but often increases water content of leaves possibly by changes in the protoplasmic structure which enable it to bind more water. Such a condition does not seem to exist in most of the wilt diseases. The loss of water-retaining capacity of the leaf cells, however, seems to be reversible in many cases, and Gottlieb (1944) envisages the continued action of a slow acting poison.

All this, regarding the role of enzymes and toxins, as well as other toxic substances in the production of disease syndromes, has remained largely speculative until very recently, having been mostly based on *in vitro* studies. The ability of the pathogens concerned to produce these substances *in vitro* could not always be correlated with the degree of their pathogenicity under natural conditions. The controversy that naturally arose out of this incompatibility led to the development of the new concept of "vivotoxin", i.e., a toxin operative *in vivo*. Although the term has been coined to define toxin action, the idea can be extended to bring within its scope the many other products ascribed a role in pathogenesis. A rigorous application of Koch's postulates, as applied for the establishment of parasitism by microorganisms would, however, seem impossible when we consider the conditions in which these factors act *in vivo*. The secretion and activity of enzymes are limited by the presence of adequate substrates and optimal conditions for their activity. Vivotoxins, on the other hand, are most probably continuously secreted and continuously inactivated or destroyed by host reactions. Hence, the establishment of their complicity in initiating certain aspects of wilt syndrome would be difficult by *in vitro* studies.

Nevertheless, recent studies by many workers have been instrumental in demonstrating the presence of pectic enzymes and metabolites of the pathogen, such as fusaric acid *in vivo* in the infected plants, together with the presence of the products of host-parasite interaction, such as ethylene and phenols. A recognition of their presence *in vivo* in the diseased plants strongly suggests the possibility of their taking part in the initiation of the array of symptoms observed during pathogenesis in wilt disease.

The similar symptoms of many wilt diseases suggest a common biochemical basis for their origin. Still, it is difficult to suggest a single comprehensive mechanism of disease initiation for all, in view of the fact that the relative importance and the presence of these factors would depend on the different host-parasite complexes.

III. "WATER IMBALANCE" IN DISEASED PLANTS

A change in the water content may, perhaps, be one of the earliest reactions of living cells to any disturbance. Most diseases are characterized by an initial increase in transpiration rate which, in itself, is sufficient to upset the water economy of a plant. There are, however, instances where a reduction of transpiration rate has been observed, e.g., bacterial wilt of cucumbers. The increase in transpiration may be brought about by various causes—physical or chemical. Changes in permeability of leaf cells are almost universally observed. The protective influences of the cuticle may sometimes be subverted by ruptures caused by fructifications of pathogens (as in rusts). The regulating influence of stomates may be altered, notably by variations in the starch-sugar equilibrium. The damage to leaf cells by diffusible toxic substances, released from the focus of infection, would contribute to the changes in leaf behaviour.

The net effect of infection, in most of the diseases studied, is a gradual decline in transpiration rate, although there may be an initial increase in water loss. This decline, in many instances, points to an inability of the infected plants to obtain enough water to maintain their turgidity and would appear to be occasioned by an impediment to absorption and conduction, resulting in an increased internal water deficit, aggravated by loss of control over transpiration.

A disturbance in auxin balance due to disease is reflected in the growth pattern of plants. The affected plants are unthrifty and exhibit poor growth. An increase in absorption may be expected to compensate for this imbalance, but this does not take place since cessation of root growth and serve root damage observed in many diseases delimit this possibility.

The conduction of water through the xylem may be variously impeded: (1) by the presence of the pathogen, (2) by the chemical activities of the pathogen resulting in formation of gums and gels, (3) by stimulation of host reactions leading to hyperplastic development of xylem parenchyma and tylose formation.

The gross metabolic changes coupled with mechanical barriers, produced in the host by parasite interaction, exert a marked influence on the water content and turnover of cells, and produce changes in the internal distribution of water in tissues and organs. The effect is initially localized, but becomes systemic in course of time. All these changes are reflected in the variety of symptoms manifested in the diseased condition, ranging from stunting and rosetting to wilting and drying up. The conditions leading to these are considered in detail in the following sections.

A. Absorption—the Dysfunction of the Root

Reduction in the efficiency of roots as absorbing organs is noticed in many diseases, such as root rots, viruses, and rusts. This results from root damage which may be produced in various ways.

1. *Root Growth is Affected.* It is generally assumed that root elongation and production of root hairs significantly increase water uptake by increasing the absorbing surface in contact with the soil. This is not necessarily true, since absorption is not always limited by the root surface in contact with the external medium, particularly if the water supply to the soil is adequate, but by internal factors, such as permeability of root tissues, the metabolic state of root cells, the capacity of the xylem to conduct water, and the gradient of the diffusion pressure deficit between the soil solution and xylem sap. In the case of diseased plants, suffering from a serious internal water deficit, however, the root density and extent are of considerable importance. A study of many examples shows that, wherever fresh production and growth of roots are initiated, the plants have a better chance of surviving disease; for instance, Mostafa (1954) demonstrated that fungal filtrates often stimulated rooting in disease-resistant varieties and he claimed this to be the mechanism of resistance in those plants. Reduction in growth is caused primarily by decrease in the water status of plants. This is because of a consequent increase in osmotic value, beyond the optimal, which reduces the intensity of vital activities.

Loss of turgor, due to decrease in water content, in the shoots is presumably accompanied by loss of turgor in the roots, resulting in injury to or destruction of root hairs and reduction in or cessation of root elongation. Suberization of the epidermis and tissue differentiation are rapid with a corresponding decrease in root elongation and consequent reduction in the proportion of root surface freely permeable to water.

Soil conditions affecting root growth include physicochemical properties of the soil, namely, texture, aeration, availability of moisture, and the total soil moisture stress. In addition to these, the presence of soil microflora, especially in the region surrounding the root—the rhizosphere—markedly influences root development. Many of these are known to synthesize and release into the soil medium metabolites which influence root growth and character. The capacity of plant roots to respond to externally applied (present?) growth factors varies in different species. Generally, only inhibitory effects (on root elongation) have been reported, since the auxin content of intact roots is normally above optimum. In the "bakanae" disease of rice, an interesting situation presents itself. The pathogen *Gibberella fujikuroi* secretes gibberellins which have growth stimulatory properties; but this effect is seen only in the shoots while root growth is inhibited. The other apparent effects of the microbial mantle, such as lowering the oxygen tension and increasing the CO_2 concentration, may also affect growth and extent of the root system. Farr (1924) presented evidence for the inhibition of root hair production in the presence of the pathogen *Fusarium lycopersici* in susceptible tomato varieties, whereas the resistant varieties behaved normally. Generally, root hair production is inhibited by lack of oxygen. It is logical to expect a lowered oxygen concentration around the roots of plants supporting quantitatively higher rhizosphere microfloras. It is well established in the case of many diseases that susceptible varieties exert a greater "rhizosphere effect" than their resistant counterparts.

2. *Roots are Injured.* An extensive destruction of the root system precedes the appearance of above ground symptoms in the case of root rot, foot rot, and other diseases caused by primitive root-infecting fungi. In many wilt diseases, however, extensive killing of the root system is postponed until after the host succumbs to the disease. Death of the root system seems to be less serious than the functional alteration in a living root system. In the former case, the osmotic barrier of the root is destroyed and water flows into the shoot by mass flow, whereas in the latter, a metabolic depression of the root cells causes a decreased absorption. Damage to tips and younger portions, rather than older parts of the roots, is more detrimental to the plant. Effects of root infections in several causes are comparable to mechanical root injuries in many respects. Crandall et. al. (1945) report that the symptoms of a loss of color of the foliage, followed by wilting in the broad leaved species and dieback in conifers, do not appear until the roots are almost completely

rotted. Heavy infection by certain cereal rusts has resulted in a rapid and severe deterioration of the root system characterized by discoloration, decrease in the number of fibrous roots, and marked loss in weight.

3. *Permeability of Root Cells is Altered.* Permeability of root cells may be affected in many ways by disease. Changes in the metabolic status of cells affect the active uptake of water by them. The changes in the osmotic gradient, which again is controlled by metabolic changes to some extent as Lundegårdh (1946) points out, may alter the osmotic movement of water through the roots.

A water deficit occurs during disease and this results in a reduction in the hydration of the protoplasm, which leads on to a depressed metabolic activity. The metabolic activity of root cells is also subject to the action of external factors, such as microbial activities and their consequences in the rhizosphere, the amount of oxygen and nature of salts present in the surrounding medium, etc. The possibility that water absorption, like salt accumulation, is not merely an equilibrium process but may involve an internal secretion peculiar to cells which are still able to grow, is suggested by the work of Bennet-Clark et. al. (1936). Internal water deficit and the presence of water depressant (antibiotics and growth factors) in the environment retard growth of root cells and inhibit synthetic processes; the latter may seriously interfere with permeability. Root cells subjected to the action of antibiotics, such as polymyxin, liberate their cell contents into the medium and it is quite reasonable to expert that metabolites which are produced in soils around the roots affect the root cells in a similar way. Moreover, any injury to root cells results in a leakage of cell contents into the medium. As a rule, healthy cells do not lose their contents, while injured cells do. It may be reasonable to presume that substance rich in nutrient value are exuded from diseased roots, thus providing a good substrates for increased microbial activity. This, in turn, could be expected to change the O_2 and CO_2 concentrations, bringing about further changes in root function.

Where root cells are killed, they lose their selective permeability. Thus, there could be an indiscriminate entry of toxic substances—of plant and microbial origin—and minerals. Many symptoms observed in aerial parts of diseased plants may be traced to such a nonselective passage of substances through the damaged root system. How toxic materials affect living cells has already been described. Plants affected by diseases such as foot rot and root rot exhibit signs of mineral deficiency. A mineral imbalance causes injury to leaves, such as mesophyll collapse observed in citrus. A pronounced ionic derangement has been noticed in the *Fusarium* wilt of cotton. All these data strongly point to the fact that there is a dysfunction of the root system resulting in selective permeability and absorption.

B. Conduction—The Dysfunction of Conductive Elements

The problem of separating the factors affecting absorption from those affecting conduction is a difficult one since both the processes are interdependent. Fluometric studies by many workers reveal that the rate of flow of water in stems of diseased plants infected with vascular parasites is reduced considerably. Many authors claim that wilting of the affected plants is due to an impairment in conduction. Power (1954) attributes the cause for wilting to tobacco affected by "black-shank" as due to the impairment of water movement through the lesions produced in the stem by *Phytophthora parasitica* var. *nicotianae*. Keyworth (1953), working on the *Verticillium* wilt of the hop, found that the severity of leaf symptoms is determined by stem invasion. The mean rate of flow is less than half as great in stem tissues of soybean plants invaded by the fungus *Cephalosporium gregatum* as in healthy tissue of comparable stem size. These workers also find an inverse relationship between the degree of browning in the vascular system and the rate of water flow. The occlusion of conductive elements in the petioles of leaves seems to be of greater significance than that in the stems inasmuch as there is little scope for circumventing this obstruction, due to lack of an alternate path for conduction and absence of secondary thickening.

The various materials that are known to contribute to the dysfunction of conductive elements include the physical presence of the organism and/or products of its chemical activity in the conductive elements. The reaction of the host tissues to the presence of the organisms in its interior also contributes a share in the production of occluding materials. A consideration of the nature, origin, and effect of these on water flow is presented in some detail below.

1. *Vessels are Choked.* (*a*) *Organisms.* A mechanical blockage of vessels has sometimes been attributed to wefts of hyphae growing freely into the lumen of the vessels and, in the case of bacterial diseases, to slimy colonies of the pathogens. Grieve (1941)) demonstrated in tomatoes and potatoes, infected with *Bacterium solanacearum*, that the progress of absorption in relation to invasion was closely similar to that of transpiration under the same conditions. Where the parasite was inocu-

lated at the stem apex, no reduction in absorption took place before the bacteria had "overrun" and "blocked" several root vessels after growing downward through the stem. The distribution of the organisms in the vascular tracts was usually found to be vertical rather than lateral and, hence, the chances of the vessels' getting blocked by the lateral spread of the pathogen are negligible and those vessels that are not infected initially remain comparatively free of mycelium. Although the presence of the organism in the lumen of vessels is noticed by many workers, the inadequacy of such an obstruction to cause the marked changes observed seems to have been realized by them. This phenomenon, however, merits consideration as a contributory factor to the reduction in water flow.

(*b*) *Enzymes.* The production of pectic enzymes by a number of vascular wilt pathogens and their role in pathogenesis have been investigated in detail in recent years. These enzymes act on the middle lamellae exposed at the pit region and liberate pectic acids and other products of partial hydrolysis of pectin, which form gels combining with metals. If the hydrolysis proceeds further, gum formation takes place. The presence of these enzymes *in vivo* has been demonstrated in some cases; for instance, in the *Fusarium* wilt of tomato and cotton. Both pectolytic and cellulolytic enzymes have been noticed in the bacterial wilt of tomato caused by *Pseudomonas solanacearum*. The fact that these enzymes are produced *in vivo* suggests their importance in tissue breakdown commonly observed in these diseases. The conditions under which they act in different cases, however, seem to vary and this determines the extent of tissue breakdown. The presence of metallic ions, particularly heavy metals and alkaline earth salts, influences their activity *in vitro*. The production and activity of these enzymes also vary with the composition of the medium in which they are built up, such as the nature of the carbon source and on the chemical nature of the substrate on which they act—the degree of esterification in the case of pectin. Gäumann and Böhni, studying the effect of nutrient solution on the production of pectinase and pectase by *Botrytis cinerea*, showed the former to be developed independent of the chemical composition of the medium, whereas the latter, which splits the methyl alcohol in pectin is largely adaptive, being produced in quantity in the presence of pectin, but only in traces without it. The substrate for pectinase is provided by the action of pectase which de-esterifies the pectin. The hydrolysis of pectic acid by pectinase is essentially zero until 45 to 50% of the bounds are hydrolyzed. This shows the synergistic action of both these enzymes on the breakdown of pectin. This especially significant when we realize that pectase is largely adaptive, depending on the presence of pectin in the substratum.

The action of the pectic enzymes on the middle lamellae of the vascular elements is to macerate the tissue and this results in the development of vascular plugs. In tomato cuttings treated with commercial pectic enzyme preparations, this effect is seen clearly. The xylem walls appear to be thinned out and the plugs formed by the enzyme action are stained by ruthenium red revealing their pectic origin. A characteristic vascular discoloration has been observed in many wilt diseases and this has been attributed to the action of parasitic enzymes such as pectin methyl esterase and to the oxidation of phenols resulting in melanoid pigments. Phenols are liberated from phenolic glycosides on hydrolysis by β-glucosidase or by shunting off the phenols from lignin formation. the lignin content of cells of infected plants was shown to be lower, thus providing evidence for this possibility. The polymerization of phenols appears to take place in the living cells of the xylem parenchyma. The substrate for fungal β-glucosidase is contained in the cells of xylem parenchyma and is made available only after a certain amount of tissue maceration by the action of pectic enzymes. Thus, the action of these pectic enzymes starts a chain of reactions in the xylem resulting in plugging and discoloration. Gäumann et. al. (1953) isolated a fraction from the fungal filtrates of *Fusarium lycopersici* responsible for vascular discoloration, the nature of which seems to be that of an enzymatic protein. The exact manner in which this brings about vascular discoloration is not known; perhaps it is one of these enzymes that takes part in the above mentioned sequence of actions. Chamberlain and McAlister (1954) report that the rate of water flow in diseased soybean stem affected by "brown stem rot" is inversely proportional to the degree of browning in the vascular system indicating the action of these enzymes, bringing about a reduction in water flow.

(*c*) *Gums and Gels.* Considerable evidence has been adduced to show that pectic degradation resulting in the release of gel and gum forming substances takes place in the conductive elements and causes considerable obstruction to the efficient translocation of water to the leaves. Dark colored gums have been found in oak wilt and these may also contribute to vascular discoloration. The presence of gum in other cases of wilts is not detectable in the early stages, but appears only after the onset of wilting; its formation is preceded by the appearance of homogeneous granular and gray material which is not preserved in fixed and stained sections, but is detectable only in living material. This happens because the dehydration in the process and the hydrophilic nature of the material do not permit its preservation. It is likely that the dehydrated granular mass lies adpressed to the xylem wall.

(*d*) *Polysaccharides.* Polysaccharides and other large molecules are known to cause wilting in tomato. These have been recognized as metabolic products of fungal and bacterial growth. It is very probable that the large polysaccharide molecules are liberated in the xylem, where these organisms are growing, in sufficient amounts to cause wilting. These substances are known to cause a mechanical blockage of vascular boundless and intermicellar spaces, thus initiating a physical wilting. If pathological wilting in plants were caused by such a mechanical blockage of vascular bundles, they should behave as plants exposed to physiological wilting from shortage of water. But this does not appear to be the case in the naturally infected plants. Moreover, if such occlusions of the intermicellar spaces of the xylem were to occur, the laminar flow of vascular stream should be greatly affected and the reversibility of the wilting action and regaining of turgidity by organs, as reported by many workers will not occur.

(*e*) *Hyperplasia.* A different type of vascular obstruction has been observed in the mosaic-infected tomato stems. In response to necrosis caused by the virus, the host cells adjacent to these necrotic spots manifest a hyperplastic development resulting in an invasion of the xylem tissue and an inwardly directed pressure. Tracheal tubes which happen to lie in the path of the hyperplastic cells are crushed and their lumen completely obliterated. In cases where a considerable proportion of the circumference of the xylem cylinder is invaded by the hyperplastic growth, it is readily conceivable that the water supply to the top might be cut off. In stems of plants infected by the crown gall organism, *Bacterium tumefaciens*, such hyperplasia is known to occur and these stems conduct considerably less water than normal.

(*f*) *Tyloses.* Occlusion of vessels by formation of tyloses resulting in impaired water flow has been noticed in many vascular diseases coincident with the appearance of wilt. The formation of tyloses is attributed to various causes such as the presence of the parasite, or a reaction to the toxic substance produced by the parasite or to a chronic water shortage. Struckmeyer et. al. (1954) found the formation of tyloses preceding the development of wilt symptoms and claim that this is the cause and not the effect of an impairment in conduction. Abundant tyloses were observed in melon plants infected with *Fusarium niveum*, their occurrence being apparently correlated with the presence, quantity, and proximity of the fungus. Powers (1954) also suggests that tyloses and gums are the main causes of obstruction of water movement in "black shank" of tobacco, rather than a result of previous rupture of the water columns. He indicates that the development of tyloses and gums in the vessels of diseased stems is induced primarily by the toxic effects of the decomposition products of the invaded cells which are not carried very far from the infection court. The formation of tyloses seems to be in no way restricted to a case of infection, since they are found in the heartwood of even healthy trees when conduction ceases. In this case the formation of these outgrowths into the xylem seems to be due to the exposure of the inner walls to air columns but injury also stimulates their production. Gum formation is sometimes attributed to the stress in the water column in the conducting vessels and it seems to be the normal defensive reaction of the host to mechanical, climatic, toxic, or parasitic stimuli. In cases where tyloses are observed, rarely are they enough to cause complete obstruction and, therefore, these cannot be considered as wholly responsible for wilting and necrosis arising out of an acute water shortage.

2. *Viscosity Changes*. The release of gums in the vascular sap may cause a change in the viscosity of the tracheal fluid. In some bacterial wilts the viscosity of the tracheal fluid in the infected plants appears to be higher and Ludwig (1952) demonstrated an inverse relation between viscosity and the rate of sap flow. However, Dimond and Waggoner claim that the xylem sap in infected tomato plant does not show considerable change in viscosity compared to healthy ones and, therefore, this factor does not appear to contribute to the reduction in water flow in the case of *Fusarium* wilt of tomato.

3. *Gas Emboli*. The formation of gas pockets in water columns of the infected xylem has sometimes been suggested as the cause of a water shortage at the tops of plants. This theory was advanced on the basis of the observations that the flax wilt organism, *Fusarium lini*, produces considerable amounts of CO_2 in cultures *in vitro*. That gas emboli appear in normal plants has not been completely overruled yet and we do not know how far this phenomenon would result in the breaking up of the cohesive forces in the water columns. If, however, the bubbles do not exceed the critical size, they are usually dissolved and the cohesion of the water column is maintained. Scholander et. al. (1955) investigated the effects of gas emboli on water conduction and found that, in spite of the presence of large volumes of gas in vine stems, rates of uptake

diminish only slightly or not at all when there is an adequate supply of water. They suggest that much of the sap flow is confined to the "finer structures between the vessels." More recent studies have indicated that any air breaks in the vessels and tracheids are confined to the units in which they occur and the transpiration stream simply moves around these obstacles in the intact units.

4. *Some Considerations on "Reduction in Water Flow"*. A consideration of the mechanism by which water is transported to the leaves through the conductive elements becomes imperative in order to assess the relative importance of the various barriers to the free flow of vascular stream. Among the theories advanced, the cohesion theory of Dixon and Jolly (1895) seems to be fairly satisfactory. However, it does not explain the phenomenon entirely. Many other theories have been advanced from time to time, notable among them being the vitalistic theories first suggested by Bose (1923) envisaging a role of pumping action in the upward transport of water by the living cortical cells. Although this was severely criticized by a number of investigators, the vitalistic theories seem to warrant some consideration. Peirce (1936) discussed the complexities of the phenomenon and concluded that it is not purely physical in nature. He observes that water is moved through the plant by physical means; however, when the living cells surrounding the vascular tissues are killed by heat, cold, or poisons, the conducting system is rendered functionless. The function of the living cells seems to consist in "maintaining a continuous but many phased water mass in the capillary body of the plant, conditioning but not compelling the ascent of sap." Handley (1939), from his experiments on the effects of low temperature on the sap ascent, believes that a chain of living cells continuous from roots to leaves is involved in the ascent of sap. Many other theories have also been proposed but they are largely based on one of these described above, i.e., physical oı involving the participation of living cells. Lundegårdh (1954) suggests, as an alternative to the cohesion hypothesis, that the forces involved in this phenomenon are the complex of capillary suction in the medium sized tracheids, electrocapillarity and other surface phenomena, while larger tracheids and vessels are air filled and play no role in conduction. The activity of living cells seems to be involved in the movement of water in trees. In contrast to the cohesion theory, Lundegårdh considers that only a small proportion of the total water in the xylem is mobile and the transpiration stream is confined to this relatively free fraction. It seems that "capillary forces in cooperation

with the absoprtive qualities of the wall substance and the osmotic imbibition of the living tissues in the stem are so successfully contributing to the maintenance of a continuous sheath of water that an ascending sap stream can be maintained by a real suction pressure originated in the transpiring leaves of moderate height perhaps even less than one atmosphere."

Viewed in the light of the present state of knowledge on the mechanism of water transport in plants, the relative importance of the various barriers considered above becomes obvious. According to the cohesion theory and that advanced by Lundegårdh involving surface forces, the states of the vessel walls assumes a great importance. By the action of the enzymes of the pathogen the composition of the vessel walls is altered, with the release of hydrophilic substances having different surface properties. The release of large molecules occluding capillaries of the intercellular and intermicellar spaces would seriously interfere with the upward movement of sap. Tyloses and mycelial fragments jutting into the water column in the vessels would cause a frictional drag and reduce the laminar flow. If, however, vessels are not actively concerned in the ascent of sap, as suggested by Lundegårdh, the presence of tyloses, gums, gels, or gas embolic assume little significance. If the part played by the living cells of the xylem is of considerable importance, their being affected by the enzymes and/or toxins would seriously interfere with conduction. Moreover, formation of tyloses, gums, or other occluding materials is neither extensive nor universal in occurrence, nor is specific to the diseased condition. Hence, the importance of these barriers would be either very great or otherwise depending on the exact nature and mechanism by which water is moved upward. The limited significance of the results accruing from fluometric studies brought forward to explain the reduction in water flow in infected stems thus becomes obvious inasmuch as they measure only the turbulent flow of water through the stem under a constant pressure head, which does not even approach the approximate condition in which water flow takes place in living plants. In the present state of knowledge, it is difficult to try to explain precisely the phenomenon of the ascent of sap and on this depends the relative importance of the barriers interfering with the flow of water.

C. Transpiration—The Dysfunction of Leaf and Stomates

Enhanced transpiration is a feature commonly accompanying pathogenesis. This is primarily due to the removal of the natural protection afforded by the cuticle and by the loss of sensitivity of stomates to respond to changes in leaf water content. Also, an increase in permeability of leaf cells responsible for the availability of water at the evaporating surface tends to increase the rate of loss of water from the leaf tissues. During pathogenesis the increased water loss from the tissues operates as a consequence of one or the other or all of the above mentioned derangements; for instance, *Phaseolus coccineous* inoculated with *Erisyphe polygoni* loses water more rapidly at night and less so during the day than healthy plants, showing that the main factor causing increased water loss is the increased permeability of the host cells. Yarkwood (1947) concludes that there is relatively little water loss from the fungus tissue itself or from mechanical openings produced by the pathogen. But in the case of rust infections of *Phaseolus* the rate of transpiration is initially about 70% of the uninoculated during the day but the rate of loss is soon doubled. The nocturnal water loss of rusted turgid bean leaves was greater than that of healthy leaves both before and after the pustules opened. During the day the turgid rusted leaves lost less water than the healthy before the pustules opened, but more afterwards. An analysis of this situation illustrates the behaviour of the leaves during disease. Initially, the rate of transpiration of the diseased leaves was lower than that of the checks and this variation was brought about by the differences in their stomatal regulation of transpiration. While the stomata on healthy leaves were wide open, those over rust pustules on the infected leaves were principally closed, thereby causing a lower rate of transpiration; but soon a condition of water deficit developed both in the control and in the rusted leaves as a result of transpiration, the stomates consequently closed; transpiration decreased and finally became cuticular. The rate of water loss from the infected leaves at this stage was higher than that of control plants, presumably due to the enhanced permeability to water of the injured cuticle over the pustules and to the incomplete closure of the stomates. In general, the rusted leaves continuously lose water more rapidly and dry out sooner without any time for the recovery of internal water deficit. Similar enhanced rates of water loss are known in the case of virus diseases also, e.g., tomato plants infected with tobacco mosaic virus and tomato spotted wilt virus. Changes in the water relations seem to be among the early reactions of the affected cells to virus proteins and the primary effect of the virus appears to be on cuticular transpiration rather than on stomatal behaviour.

1. *Leaf Surface is Reduced*. Extensive reduction in leaf area is brought about by various aberrant growth phenomena in leaves due to attack by

pathogens. Leaf spot diseases considerably reduce the proportion of healthy leaf cells essential for the development of a suction force adequate to cause a flow of water into the leaves. Other deformities, such as leaf roll, leaf curl, and little leaf, caused by various viruses which induce imbalanced or inhibited growth and expansion of the lamina, also act in the same manner. Defoliation of leaves if a feature accompanying many leaf infections and this seriously interferes with the uptake of water by plants. In addition to reducing the suction force developed by the leaves, such injuries to leaves lead to many far reaching consequences, such as a derangement in the metabolism, starvation due to reduction in the assimilatory tissue, leading to a progressive degeneration and a disturbed auxin balance.

In addition to the external deformities, certain anatomical changes in the affected leaves are brought about as a result of infection in certain cases. Peach leaves infected by *Taphrina deformans* exhibit such a deformity resulting in thickening of leaves. The palisade cells multiply and lose their usual elongated shape and become isodiametric. Structural changes have been noticed in certain virus diseases also, e.g., potato leaf roll, tobacco leaf curl, little leaf of brinjal, etc. Mesophyll deformation has been observed in cranberries infected by *Exobasidium oxycocci*. In these cases the intercellular spaces are completely obliterated and the leaf becomes stiff. Stiffening is sometimes caused by toxins as in the case of lycomarasmin action on tomato leaves. The exact manner in which this effect is brought about by toxins is not known.

2. *Leaf Cells are Damaged.* Damaged to leaf cells is caused by both the physical and the chemical action of the parasite.

(*a*) *Necrosis.* Infection takes place either through the stomates or through the cuticle. In the latter case, a rupture is caused either by the pressure of the growing germ tube or by the dissolving action of the enzymes that are secreted by the tip of the advancing germ tube. The growth of the organism inside the leaf tissue may be intercellular or intracellular; in either case, the spread of the organism involves considerable damage to leaf cells. Pectolytic and cellulolytic enzymes are secreted by the spreading mycelium dissolving the cell wall material. The action of these enzymes on the walls of the leaf cells seems to bring about maceration and rotting, culminating in killing of the tissues. In other cases, although such extensive damage to cell walls is not noticed, the protoplasts are killed and their death is accompanied by accumulation of phenolic substances which turn brown on aging and impart that hue to the dead cells. Considerable change in the constitution of the protoplast is brought about by the action of the parasite resulting in various chemical reactions interfering with the utilization of inorganic phosphates. The excretion of phosphorus in the intercellular spaces by cells which normally retain this element is accompanied by an internal secretion of phenolic substances forming coacervates. The coacervate formation affects the metabolism of the cell and the distribution of nucleotides and phosphoproteids with the resultant decompensation of respiration. The latter process may assume various degrees of severity, a mild form permitting the survival of the host cells, while severe damage results in the development of characteristic, hypersensitive, necrotic lesions. In the case of "wildfire" disease of tobacco the affected cells are starved and become devoid of starch and sugars. Oil droplets accumulate and the plastids are disintegrated. Fat soluble yellow carotene pigments, on destruction of chlorophyll, become evident in the yellow halo produced around the spots. The action of the toxin, reported to be functional in this disease, in through competitive inhibition, preventing the utilization of L-methionine.

In the case of wilt diseases a different types of injury is discernible. The earliest visual symptom of veinclearing is observed in the case of *Fusarium* wilt of cotton and tomato. In cotton leaves showing veinclearing, histologic examination reveals a disintegration of the plastids in the cells adjoining the veins. It has been suggested that this is caused by the translocated toxins that permeate the leaf tissues through the veins. Similar symptoms have been observed in most virus diseases but the clearing of veins in these instances appears to be due to a suppression of formation of plastids rather than their disintegration. Long before the appearance of definite veinclearing symptoms, the veins and venalities in the leaves exhibit a characteristic fluorescence when viewed under ultraviolet light. As stated earlier, the living cells are exposed to the action of fungal β-glucosidases which act on phenolic glycosides inside the cells, therating conjugated phenols. An injury caused by such liberated phenols results in increased transpiration.

It becomes evident that the changes taking place in the leaves would result not only in increased loss of water due to destruction of forces that retain water against evapo-transpiration, but also in the progressive deterioration of the suction force normally developed pin them which, then transmitted to the roots, facilitates absorption.

(*b*) *Permeability Changes.* Depending on the extent of injury to cells, an increase or a decrease in permeability to water occurs. When injury is severe enough to result in death of cells, a marked increase in permeability to both water and solutes occurs. Permeability of protoplasmic membranes is maintained by continued expenditure of energy provided by respiration and any change in respiration affects permeability accordingly. Factors that inhibit growth also inhibit the uptake of water by cells. In the case of plants infected by various parasites, depending on the type of injury each produces, corresponding changes occur in cell permeability to water. An increase in permeability of bean leaves infected by powdery mildew and rust was reported by Yarkwood (1947). Increased permeability may be found due to a direct action on the plasma membrane, resulting in leakage of cell contents. Such a mechanism has been proposed for the action of lycomarasmin, one of the wilt toxins produced by *Fusarium lycopersici*. This results in the destruction of osmoregulatory functions of the protoplasts, causing a release of water from the cells. Fusaric acid, another wilt toxin identified in the diseased cotton and tomato plants, increases the permeability of cells to water at low concentrations, although a decrease is observed as the concentration of the toxin increases. Many of the other fungal toxins/antibiotics, e.g., alternaric acid, penicilic acid, patulin, streptomycin, etc., have been shown to impair permeability even at very low concentrations.

The exact manner in which these diverse groups of substances bring about the changes in permeability is not yet known. However, it may be possible to imagine that they act upon one or more of the permeability barriers of a cell, namely: (1) the cell wall, (2) the protoplasmic membrane or plasmalemma, (3) the protoplasts itself, and (4) the tonoplast. For instance, cellulolytic and pectolytic enzymes produced locally or transported through the transpiration stream might bring about dissolution of the walls of leaf cells, thereby eliminating or weakening one of the barriers, the effective functioning of which is necessary for maintenance of turgor. The implication suggested here is that when protoplasts are enclosed by rigid walls, their permeability may be decreased or increased, depending on their turgidity. Polysaccharides and other large molecular carbohydrates, either produced by the parasite or due to breakdown of starch resulting from a condition of water deficit in leaves, may also affect the permeability of leaf cells by blocking the intermicellar capillaries in the cell walls and cause an irreversible impairment to movement of water. The action of substances on the protoplasts causing a change in permeability and water retaining capacity seems to be brought about in many ways, such as an interference with the rates of metabolism that maintain the colloido-chemical properties of the protoplasm, the protoplasmic membranes, synthesis of carrier ions, etc.

(*c*) *Osmotic Changes.* Changes in osmotic pressure result from variations in water content and from variations in solute concentration. The continued loss of water through transpiration, coupled with an inadequate supply of water, produces an increasing water deficit which initiates a chain of reactions, in addition to causing an increase in osmotic pressure. Hydrolytic processes, rather than synthetic processes are stimulated, and this leads to the formation of osmotically active solutes; the accumulation of these in the cells, due to changes in respiratory activity, contributes to an enormous increase in the osmotic pressure of sap in the cells. In some cases, as in the *Fusarium* wilt of cotton, an ionic imbalance in the leaf cells as shown by spectrochemical studies and conductivity studies, has been noticed. The net effect of infection in these cases appears to be an imbalance between monovalent and divalent cations, as indicated by the loss of potassium and accumulation of calcium, manganese, magnesium, etc., such an imbalance is known to influence the permeability of cells to water and, consequently, their osmotic pressure. An ionic imbalance is also known to produce protoplasmic shrinkage, resulting in a water loss and this suggests that the toxemic condition produced by many of the wilt toxins probably is preceded by an ionic imbalance, leading to the above changes.

(*d*) *Water Status of Leaf Tissue.* It will be evident from an examination of the sequence of events described above that more water deficit progressively increases, culminating in the death of cells. A study of the metabolic patterns in the leaves of plants infected with various pathogens reveals the existence of a serious internal water deficit. For instance, an increase in the amount of reducing sugars, nonprotein nitrogen, conductivity of the sap, and respiration has been observed in the *Fusarium* wilt of cotton. This reflects a condition of water deficit.

3. *Pathological Loss of Water.* The loss of water from diseased plants, which exceeds water intake, is brought about by the disruption of cuticle, dysfunction of stomata, and increased permeability of leaf cells by the action of the pathogen. This excess leads to a progressive decrease in fresh weight, not to mention the other important changes in the vital functions of the plant. Transpiration is essentially a passive process and in the absence of controlling influences, such as those exerted by cuticle, stomata, and the diffusion pressure deficit of leaf cells, would continue

according to the evaporation power of the atmosphere. An excess of transpiration over absorption is a phenomenon of daily occurrence in normal healthy plants when conditions favor this process, but soon the controlling effect of the above mentioned factors comes into play and checks its continuation. This does not appear to happen in the diseased plant. However, the loss of turgor and consequent wilting under pathogenesis appear to be brought about, not only by the excess of water loss, but by changes leading to the destruction of osmoregulatory mechanism of the leaf cells and, hence, the rate of transpiration does not seem to indicate all the aspects of pathological wilting. This fact becomes obvious from the experiments where wilting is induced in excised shoots treated with toxin and placed in a saturated atmosphere where the rate of transpiration is negligibly low. It would thus appear that it is not so much the gain of water by absorption or loss due to transpiration that it of vital importance to plants but the manner in which the plant, at the level of individual cells and tissues, is able to maintain its hydration at the optimum which controls the various metabolic functions and, in turn, is controlled by them.

IV. SYNTHESIS

From a comparative study of various diseases it is clear that a disturbance of the normal water balance occurs at some stage or other, although it is not possible to say whether it is the cause or consequence of the other phenomena associated with pathogenesis. The general condition of water imbalance in diseased plants is associated with a derangement in its absorption, transport, and/or transpiration and is often accompanied by a disturbance to the various other basic processes, such as carbohydrate and nitrogen metabolism, respiration, and mineral up-take. A study of the metabolic processes that contribute to the maintenance of a normal water balance would, therefore, be necessary in order to understand the principles involved. For instance, the changes in the osmotic pressure of cells and tissues, contributing to the maintenance of a proper gradient facilitating absorption of water and its internal redistribution to various crgans to maintain their hydration at an optimum level, depend on the balance between hydrolytic and synthetic processes. These processes, in turn, are influenced in no small measure by the ionic balance in the tissues. It would thus appear that a disturbance in any one of these would initiate a chain of reactions and would lead to an imbalance of various key metabolic functions, and it is very likely that disease arises out of a disturbance to any one of these interrelated processes.

The present state of knowledge, in the absence of a complete picture of the simultaneous changes in all these processes which are linked closely to one another, would at best indicate a correlation rather than a causal relation between the various phenomena accompanying a disease. It is implicit that one should exercise utmost caution in attempting to define the cause of a particular derangement, in the absence of collateral studies of various interrelated processes. It looks as if the constitution of the cell is such that its different reactions would be funneled in a way that would primarily affect its water balance. It is, therefore, reasonable to presume that the primary effect of a disease in on the water status of the cells, which reflects in the various derangements in metabolic functions, leading to a progressive degeneration of vital activities, culminating in death. Thus, it seems to be verily true that "the fire of life burns in water."

3

EFFECT ON RESPIRATION

Among the many metabolic changes observed in diseased plant tissue, alteration of respiration is one of the most fascinating subjects. From the biochemical information obtained on this phenomenon, one can gain valuable information about the mechanisms controlling metabolism in plants on one hand, and an elucidation of the underlying principles of the host-pathogen relationship on the other. Respiration occupies a central part of metabolism by providing energy to support cellular processes and is important since the energy from respiratory metabolism is used by the host to carry out its responsive reactions against pathogens. An increase in the respiratory rate is a typical feature of the metabolism of infected plants and has been studied very extensively. Sometimes, when a pathogenic attack on a plant leads to a degenerated condition, the host's rate of cellular catabolism may decline and come to a standstill. Thus, eventually respiration will stop.

In healthy host cells, metabolism progresses in a well-balanced manner. Since all organisms have some flexibility in controlling their metabolism, it can be assumed that they are able to adjust their metabolism to a new circumstance, so that with a slight environmental modification the over-all reaction will not be altered significantly. However, with a more drastic change of circumstance, such as the presence of a pathogen, the formation of new metabolic pathways may be observed. We can visualize this situation in the sequence below in which A is normally concerted through intermediate B to C. If the velocity of reaction K_1 is equal to that of K_2, then A is connected to C without accumulation of B. However, if K_1 is proceeding much faster than K_2 or the velocity of K_2 is negligible, then B will accumulate and sometimes new reaction

$$A \xrightarrow{K_1} B \xrightarrow{K_2} C, \quad B \xrightarrow{K_3} X, \quad B \xrightarrow{K_4} Y$$

sequences may be induced. K_3 and K_4 indicate those alternative or shunt pathways leading to new abnormal metabolites X and Y. This type of metabolic alteration has been observed in diseased tissues or organs. Many factors affecting this metabolic alteration and varied reactions of host tissue will be surveyed.

In this chapter, four aspects of the alteration of respiration in plant tissues, infected by pathogenic microorganisms, will be discussed. First, we would like to describe the respiratory increase of infected plants. Several biochemical and physiological hypotheses have been put forward to explain it and the information pertaining to this condition in infected plants will be discussed in detail. The Pasteur effect will be discussed in connection with this, since some authors have obtained experimental evidence showing a relationship between the respiratory increase and an inhibition of the Pasteur effect in diseased plant tissue. The possibility that an alternate respiratory enzyme system is activated and participating in the respiratory increases will be considered together with the facts concerning the accumulation of some substances. Second, we shall discuss the H.M.P. pathway. Since this is recently elucidated pathway for carbohydrate metabolism, few experimental results showing the operation of this pathway in the respiratory pattern of infected plants have been reported. Third, we shall discuss some biochemical phenomena accompanying the respiratory change, such as synthesis of aromatic compounds, metabolism of amino acids, auxin, and lipids, and the physiological function of these phenomena in the host-pathogen relationship. Finally, we shall discuss the relationship between respiratory metabolism of infected host tissue and its defense mechanisms. Unfortunately, this is mostly a speculative field, but it might not be too presumptuous to say that the purpose of the research described in the preceding parts is to elucidate the defense mechanism of the host.

In general, we have tried to point out the important problems for future study rather than describing the scattered reports in this field. Although relatively little work has been done on the physiological and biochemical aspects of plant disease, several articles are helpful in understanding the physiological problems of host-pathogen relationships. The enzymological aspects of plant diseases have been discussed by Farkas and Kirâly (1958); their article contains a considerable number of reports on the oxidative enzymes of infected plants. The alteration of host respiration and its relation to resistance was discussed in a symposium on the "Mechanism of Resistance in Higher Plants", held in Japan in 1956.

I. RESPIRATORY INCREASE

A. Mechanisms Controlling Respiration

Respiratory increase is not a phenomenon uniquely associated with a few plants injured by a specific pathogen, but a kind of general responsive reaction of the plant tissue attached by pathogenic microorganisms including fungi, bacteria, and viruses. As will be seen in Table 3.1, it has been observed in white potatoes infected by *Penicillium* spp., *Ceratostomella fimbriata, Phytophthora infestans*, and in sweet potatoes attacked by *Rhizopus tritici, Ceratostomella fimbriata*, or *Helicobasidium mompa.* A similar responsive reaction has also been observed in rice and wheat plants infected by several fungi. It should be emphasized that both obligate and facultative parasites are able to evoke this respiratory increase in host tissue; furthermore, simple chemical treatments or mechanical stimulation is also able to induce an increase in the respiratory rate of plants.

TABLE 3.1
Examples of the Respiratory Increase in Infected Plants

Host	*Pathogen*	*Enzyme or enzyme system studied*
	Fungus Pathogens	
Barley leaves	*Erysiphe graminis*	
Barley leaves	*Erysiphe graminis*	H.M.P. pathway
Cabbage leaves	*Botrytis cinerea*	Peroxidase
Cabbage leaves	*Botrytis cinerea*	Ascorbic acid oxidase
		Cytochrome oxidase
		Peroxidase
		Flavoprotein enzyme
Cotton stem	*Fusarium* spp.	Peroxidase
Cotton stem	*Fusarium vasinfectum*	—
Potato tuber	*Penicillium* spp.	Catalase
		Oxidases
Potato tuber	*Ceratostomella fimbriata*	—

Host	*Pathogen*	*Enzyme or enzyme system studied*
Potato tuber	*Phytophthora infestans*	Polyphenol oxidase
Potato tuber	*Phytophthora infestans*	Polyphenol oxidase
Rice plant leaves	*Cochliolus myyabeanus*	Polyphenol oxidase
Rice plant leaves	*Piricularia oryzae*	Flavoprotein enzyme
Safflower hypocotyl	*Puccinia carthami*	H.M.P. pathway
Sweet potato root	*Rhizopus trilici*	—
Sweet potato root	*Ceratostomella fimbriata*	Cytochrome oxidase Peroxidase Polyphenol oxidase
Sweet potato root	*Helicobasidium mompa*	Polyphenol oxidase
Wheat leaves	*Erysiphe graminis*	—
Wheat leaves	*Erysiphe graminis*	—
Wheat leaves	*Puccinia graminis*	H.M.P. pathway
Wheat leaves	*Puccinia graminis*	Ascorbic acid oxidase
	Bacterial Pathogen	
Tomato hypocotyl	*Agrobacterium tumefaciens*	Terminal oxidase
	Virus Pathogens	
Tobacco leaves	Tobacco mosaic virus	—
Tobacco leaves	Tobacco mosaic virus	Catalase Peroxidase
Tobacco leaves	Tobacco etch virus	—
Bean leaves	Tobacco mosaic virus	—
Nicotiana gluttnosa	Tobacco mosaic virus	—

It is necessary to define the term "respiratory increase" which appears frequently in the following discussion. In principle, oxygen uptake permit weight or volume of infected plant tissue is measured and compared to that of an uninfected plant. In most cases, it is determined for the healthy tissue adjacent to injured tissue; this is then compared to the respiratory rate for the same tissue of an uninoculated control plant. However, in the case of leave diseases such as wheat rust, the magnitude of the respiratory rate is often determined by using tissues including both pathogen and infected host tissue, because it is difficult to separate the uninfected from the infected region. The experiments of several workers have shown that the enhancement of the respiratory rate of infected plants can be attributed in part to an independent phenomenon exhibited by host tissue, rather than being due solely to the respiration of the invading microorganism.

In recent years, biochemical investigations of this interesting phenomenon have increasingly received the attention of researchers. Before describing these, it would be valuable to have a general picture from a biochemical point of view of the mechanisms controlling respiration. An excellent review dealing with this subject appeared recently. In this article, the author discussed the biochemical principle of the control of rate of respiration, and also tried to interpret several examples of respiratory increase observed in plant tissue by means of a control mechanism. There exist many factors which control the respiratory metabolism of the living cell. They are, for example, concentration of enzymes, substrates, and cofactors, influence of pH, temperature, redox potential, effect of inhibitors; the effect of inhibitors and concentrations of enzymes, substrates, and cofactors will be covered in this section. However, since specific reports are lacking regarding the rest of the topics, they will not be discussed here.

It is believed that the over-all rate of respiration of living organisms is governed by the concentrations of the "pace-makers" in tissue. Recent biochemical studies have shown that these "pace-makers" are the concentrations of ADP and Pi. This mechanism will be considered as follows: in the steps of the phosphorylation reaction accompanying the respiratory chain oxidation of DPNH, 3 mole of ATP are generated. Although there is no complete agreement regarding the sites of ATP synthesis from ADP and Pi in the respiratory chain, present knowledge concerning this point will be shown diagrammatically in Fig. 3.1. Since oxidative and phosphorylative reactions are not separable but closely

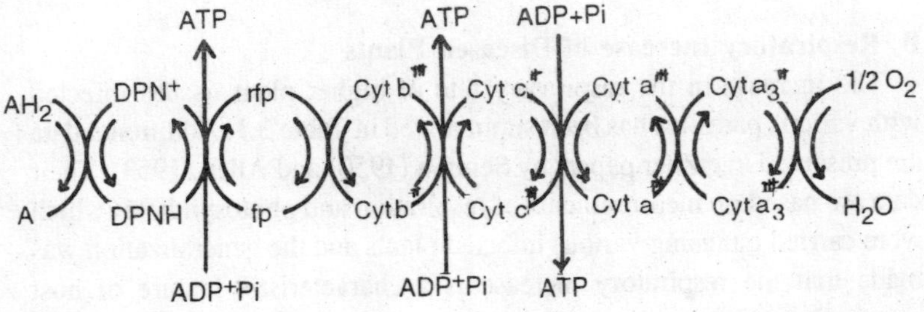

Fig. 3.1. Sites of ATP synthesis from ADP and Pi in the respiratory chain oxidation.

connected to each other, continuous flow or supply of ADP and Pi is a prerequisite for the smooth progress of respiration. Oxidative phosphorylation without continual use of the synthesized high energy phosphate may result in a shortage of phosphate acceptor (ADP) and an eventual standstill of respiration will occur. As would be expected, this situation may be overcome by the regeneration or further addition of phosphate acceptor. There is some difficulty in obtaining conclusive evidence that this mechanism is actually functioning in intact tissue; however, based on experiments using a cell-free system, this is presumed to be the case. An intimate relationship between the concentration of phosphate acceptor and the rate of respiration was first clearly established by experiments using rat liver mitochondria, thereafter, several workers using different materials have carried out the same sort of experiment with a similar result. Now, provided the turnover rate of the ATP breakdown reaction—resulting in the generation of ADP—will regulate the rate of oxygen uptake in tissue respiration both *in vivo* and *in vitro*, any system accelerating the breakdown of ATP will increase the rate of respiration. There are several reaction sequences which will utilize and/or break down ATP.

$$ATP + R \rightarrow ADP + R\text{–}P \qquad \text{(R : phosphate acceptor)}$$

Therefore, our primary interest is to look into the exact mechanisms of these ATP-utilizing reactions in infected plants.

However, the fact should be emphasized that, in some cases, concentrations of enzymes and substrates can influence the over-all rate of respiration. For instance, a shortage of carbohydrate will suppress the respiratory oxidation when the infected host tissue is seriously degenerated. Another example is the case in which synthesis of respiratory enzymes increases the rate of cellular respiration.

B. Respiratory Increase of Diseased Plants

The increase in the respiratory rate of higher plant tissues infected with various parasites has been summarized in Table 3.1. Additional data are presented in earlier papers by Sempio (1950) and Allen (1953). Their data are based on measurements of respiration and photosynthesis which were carried out using various infected plants and the generalization was made that the respiratory increase is a characteristic feature of host metabolism. On the contrary, recent work has been much more concerned with the elucidation of the mechanism of the respiratory increase, and is somewhat more biochemical and enzymatic in nature. These newer observations have been selected in Table 3.1.

Research on the respiratory increase of wheat leaves infected with powdery mildew (*Erysiphe graminis*) comprised one of the earliest reports. Allen (1953) implied that the respiratory increase of the rusted wheat leaves could be caused by the action of a phytopathogenic toxin which uncouples the oxidative phosphorylation of the host tissue. His excellent article is, so to speak, a milestone in recent biochemical and physiological studies on the respiratory increase in host tissue, and evidently has stimulated many later workers. Some workers have carried out their work to confirm Allen's thought, while others have developed their work starting from his hypothesis. It was then proposed that the active metabolism in the host leading to an increase in the turnover rate of ATP breakdown might also cause the respiratory increase. In his second article, Allen (1954) also extended his idea and suggested that the synthetic processes, accelerated in plants when infected with obligate parasites, may cause the augmented respiration. He did not obtain firm experimental evidence to prove his first hypothesis, but came to this conclusion based on an analysis of Sempio's work. As will be shown, the proposed action of a toxin is like that of DNP which by its uncoupling action will accelerate the breakdown of ATP in host tissue. Consequently, respiratory increase may result. Millerd and Scott (1956) published two papers concerning the changes in the respiratory rate observed in barley leaves infected with powdery mildew caused by *Erysiphe graminis*. They prepared a crude extract from the infected barley leaves and observed a slight respiratory increase in noninfected leaves when the extract was added. Partially purified extracts also showed a similar effect. Subsequent attempts to obtain evidence of toxin production in rusted wheat and rusted safflower have been unsuccessful. However, the Hungarian group has just obtained experimental data to support the original hypothesis of a toxin mechanism proposed by Allen. That is, in the leaf tissues of the rust-infected wheat, acid-soluble organic phosphate decreased, concomitant with an increase in Pi.

One can consider many possibilities for the biochemical action of a phytopathogenic toxin, of which uncoupling would be just one. Let us consider some other aspects of toxin effect in relation to host respiration. The following are examples of toxins or toxin containing extracts which have been isolated from the culture media of pathogens and/or from infected plants. Evaluation of their pathological relation should be made very carefully, as mentioned by several authors, since phytopathogenic toxins isolated from the cultural filtrates of pathogens have not always been found in infected plants.

The recent work of Wheeler et. al. (1958) on the mode of action of a toxin—victorin—produced by *Helminthosporium victoriae* is of extreme interest. These workers examined the effect of victorin on the respiration of tissues of oat varieties. Treatment of root, shoot, or leaf tissues of intact seedlings with victorin resulted in twofold to fivefold increases in the rate of respiration. This response of the plant was detectable after 2 hours and it reached a maximum after 4 to 10 hours. With susceptible oat varieties, victorin caused a respiratory increase 2 to 3 times greater than that caused by DNP. The important fact, however, is that victorin had no effect on the respiration of oat varieties not susceptible to *Helminthosporium victoriae*, whereas DNP affected both types of varieties. Another interesting observation is that homogenates of tissues treated with victorin showed threefold to fourfold increases in the rate of oxidation of ascorbic acid but no corresponding increases in activities of the cytochrome or pholyphenol oxidase systems. It appears that the effect of victorin is not due to the uncoupling action but is more complicated.

Tamari and his group studied the rice plant disease caused by *Piricularia oryzae* and succeeded in isolating two toxins from both the culture medium and severely injured plant tissue. One of them, α-picolinic acid, has a chemical structure similar to that of fusaric acid; the second substance found was named piricularin. Fusaric acid was first isolated from the culture medium of *Gibberella fujikurio* and identified by Yabuta and his co-workers (1934). Both α- picolinic acid and fusaric acid inhibit porphyrin-containing enzymes by a chelating action. The chemical structure of piricularin has not been conclusively demonstrated, although it has been isolated in a crystalline form. From his observations on the physiological properties of piricularin, Tamari has stated that the mode of action of this toxins is as follows. Polyphenol oxidase which might function as a formal terminal oxidase in the rice plant is inhibited by piricularin, because it forms a complex with the substrate, chlorogenic acid. Thus, the metabolism of rice plant will consequently be damaged. Relatively high concentrations of piricularin inhibit the respiration of the rice plant (1 to 4×10^{-4} gm per liter). However, the respiration of plants treated with very dilute solutions (1×10^{-6} gm per liter) of this substance shows an increase of 10 to 15%, presumably because of the response of the plant to the toxin, and not because of an uncoupling action of the toxin. As will be discussed in more detail later, the assumption that polyphenol oxidase acts as a terminal oxidase in rice plant should be considered with caution.

Paquin and Waygood (1957) have examined the effect of toxins of *Fusarium oxysporum* f. *lycopersici* on the enzymatic activity of mitochondria in the tomato hypocotyl. At a concentration of 10^{-2} *M* both lycomarasmin and fusaric acid inhibited the activities of succinic oxidase and cytochrome oxidase, but the inhibition could be completely reversed by the addition of a catalytic amount of cytochrome c. They postulated that the possible effect of lycomarasmin on mitochondria may be in part an alteration of their structural integrity, thus leading to a diffusion of cytochrome c from the particles. They also discussed these observations in connection with the tomato wilt symptoms caused by lycomarasmin. Their opinion is somewhat different from the earlier idea that lycomarasmin destroys the selective permeability of the plasma membrane of the host cell. Fusaric acid has already been shown to have an inhibitory action on the respiration of tomato tissue. It appears that frequently a toxin can act as a respiratory inhibitor rather than as an uncoupling agent.

Arzichowskaja (1946) isolated a crude toxin from the cultural medium of *Botrytis cinerea* and showed that the respiration of both infected and healthy cabbage leaves infiltrated with the toxin was augmented. In 1942 it was reported that substances isolated from the culture medium and mycelium of *Gibberella saubinetii* stimulated the respiration of potato tuber. The author suggested that this substance was likely to be pantothenic acid. Following a similar procedure, Indian workers examined the influence of cultural filtrates of several species of *Fusarium* on the respiration of cotton and observed an increase or decrease of tissue respiration depending on the species of *Fusarium* used. Since several compounds, such as organic acids, amino acids, and vitamins, present in the cultural filtrates, have been shown to cause an increase in the respiratory rate of the plants, they assumed that these substances are probably playing a role in stimulating host respiration. In this case, however, the compounds are very likely accelerating the respiratory rate by providing substrates for respiratory oxidation by the plant.

As a guide for future study on the pathogenicity of microorganisms from the standpoint of respiratory alteration, it is worthwhile to refer to the elegant work of Pappenheimer (1954) on the diphtheria toxin. He studied the biosynthesis of the toxin in *Corynebacterium diphtheriae* and obtained evidence that it may be cytochrome component of this bacteria. From his further study on its inhibitory effect on the respiratory system of silkworm, it appears that the pharmacological action of the toxin may be attributable to a blockage of cytochrome synthesis in susceptible animals. According to this mechanism, the toxin would undoubtedly disturb the cell economy of the host either by the total disturbance of the respiratory pattern or by inhibiting oxidative phosphorylation.

It is conceivable that some metabolic substances formed in the infected host tissue may have a biological action more or less similar to that of a toxin produced by pathogens and functioning in the respiratory increase of the host. This possibility has been examined in the case of sweet potato black rot, a disease caused by *Ceratostomella fimbriata*. It has been found that ipomeamarone, a substance produced as an abnormal metabolite of infected host tissue (see Fig. 3.7), has a stimulating action on the respiration of sweet potato when used in low concentrations (30–50% increase at 4×10^{-3} gm per liter). Higher concentrations, on the other hand, lower the respiration of sweet potato (50% inhibition at 4×10^{-2} gm per liter). Since ipomeamarone exerts an uncoupling action on the oxidative phosphorylation of sweet potato infected with black rot fungus could be explained by means of an uncoupling action of ipomeamarone. However, analysis for the presence of ipomeamarone has shown that this substance can hardly be detected in the healthy portion of an infected sweet potato, a region where respiratory increase is observed. Therefore, it is difficult to believe that the major portion of the respiratory increase is attributable to the uncoupling action of ipomeamarone. We would rather consider the action of this substance as a side effect in the respiratory increase. The main function of ipomeamarone in the hostpathogen relationship may lie in its fungitoxic nature and necrotic action against the host. The quantity of ipomeamarone accumulating in the infected part of sweet potato is around 10 gm per gram fresh weight in most cases. This concentration is sufficient to produce an inhibitory effect on both the host and the pathogen (Table 3.2).

Furthermore, there are two other possibilities bearing on the uncoupling type respiration in infected plants. Since activity of oxidative phosphorylation is closely associated with the biochemical structure of mitochondria, and since in some pathological circumstances plant mitochondria might be destroyed, respiratory oxidation would not be coupled to phosphorylation. A second possibility is the operation of an alternative respiratory pathway other than the TCA cycle-cytochrome system which does not link to the phosphorylation reaction. Experimental findings concerning the activation of this system are discussed in the following section; however, its relation to oxidative phosphorylation has not been studied thoroughly.

TABLE 3.2

Respiration and Formation of Polyphenols and Ipomeamarone in Sweet Potatoes Infected with *Ceratostomella fimbriata*[a]

Variety	*Resistance*[c]	*Respiration*[b]		*Polyphenols (mg./gm. fresh weight of tissue)*	*Ipomeamarone (mg./gm. fresh weight of tissue)*
		Oxygen uptake (mu./hr.)	*% Increase over control*		
Norin No. 10	+++	173	70	Not analysed	Not analysed
Norin No. 1	++	140	87	4.6	18.9
Chugoku No. 5	++	160	60	3.9	19.8
Nakamurasaki	++	144	55	4.1	14.5
Okinawa No. 100	++	140	118	3.2	9.4
Kanto No. 35	+	134	56	3.6	8.9
Kenroku	+	132	142	4.7	9.4
Suigen	–	123	45	4.3	9.8
Norin No. 5	–	88	67	4.4	5.6

[a] Sweet potatoes were sliced (1–2 cm thick), and some of the slices were inoculated with a spore suspension of *Ceratostomella fimbriata*: the rest served as a control sample. They were subjected to further analysis.

[b] The number of plus signs represents the degree of resistance of sweet potato against the pathogen in a decreasing order. The minus sign represents a susceptible variety.

[c] After 48 hrs inoculation of the fungus at 25°C, 20 slices, 0.5 mm thick and 7 mm in diameter each, were prepared from the healthy tissue adjacent to the infected region, and respiratory oxidation was measured manometrically at 30°C. Sound tissue was obtained from the control sample and its respiratory activity was compared with that of the infected one.

Another explanation for the mechanism of respiratory increase may be found in the active metabolism accelerating the ATP-utilizing reactions of host tissue, as suggested by several authors. Many anabolic processes requiring energy undoubtedly accelerate ATP breakdown and consequently increase the respiratory rate. In support of the above mechanism, many characteristic features indicating activation of the metabolism of the host have been reported such as (1) accumulation, mobilization, and synthesis of phosphorus and carbon compounds,

(2) growth of the host tissue, (3) synthesis of proteins including activation of enzymes and enzyme systems, and (4) increase in protoplasmic work, such as protoplasmic streaming (cyclosis). The over-all picture of these events is diagrammatically shown in Fig. 3.2.

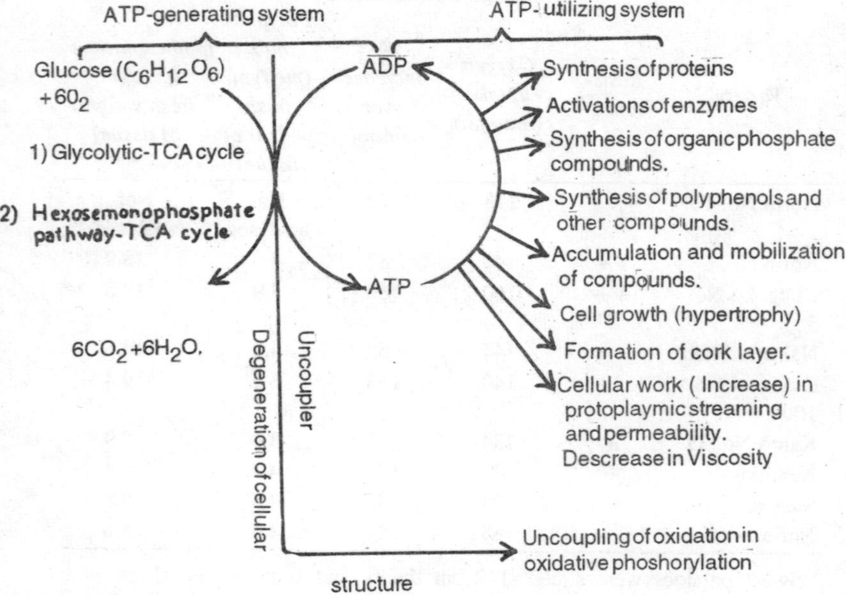

Fig. 3.2. Relationship between respiratory increase and ATP-utilizing systems in infected plants.

The first case has probably been studied most extensively and many articles on this subject using radioisotopic techniques have been published. Several workers have observed the accumulation of P^{32} in the wheat leaves infected with rust. Yarwood and Jacobson (1950) have also reported selective absorption of S^{35} by the infected leaves. Recently, this problem has been investigated by Canadian workers on a large scale. They have examined the assimilation of $C^{14}O_2$, the uptake and the distribution of P^{32} and other radioactive organic substances—including carbohydrates, organic acids, and amino acids around the spots on the cereal leaves infected with rust and mildew. The accumulation of glucose-1-C^{14} occurs in the host tissue at the site of the fungus colonies and parallels the increase of respiration. Interestingly, the accumulation was not observed in the tissue of the host killed by the pathogens. This accumulation was also inhibited by anaerobiosis and by some phosphorylative and respiratory inhibitors like DNP, azide, or H_2S. The authors came to the conclusion that the accumulation and mobilization of carbohydrates and organic and amino acids under pathological conditions are active processes closely connected with the respiration of the host. Net synthesis of organic phosphorus compounds has been observed in sweet potato infected with either *Ceratostomella fimbriata* or *Helicobasidium mompa.*

The work of Daly and Sayre (1957) is of interest for the discussion of case number 2. They have studied the relationship between cellular growth of the host and the increase in the respiratory rate of safflower infected with *Puccinia carthami*. During the vegetative development of the pathogen, the diseased hypocotyl elongated approximately twice as rapidly as did the healthy hypocotyl and showed a greater absolute length and weight with a concomitant doubling in the host respiration. They concluded that the abnormal growth of the rusted host demands energy, thus accelerating ATP breakdown and eventually enhancing the respiratory rate of the host. This assumption eliminates the need for an uncoupling mechanism by a toxin. Accordingly, they failed to get evidence of toxin production. These authors emphasized the importance of study on the hormonal control of both the growth and the respiratory metabolism of the host. A recent study on the physiology of host-pathogen relationships using rusted wheat deserves detailed description. A few of their experimental results are shown in Figs. 3.3 and 3.4. In the case of susceptible species (Little Club), but not of resistant species (Khapli), of wheat infected with rust, hypertrophy was observed with a respiratory increase indicating the unbalanced elongation of the host cells under pathogenic infection. Perhaps we can invoke a mechanism similar to that of Daly and Sayre for a possible explanation of the respiratory increase in rusted wheat. Localized hypertrophy is frequently observed in hosts infected by obligate parasites.

There are several reports on protein synthesis and on the activation of enzyme systems in infected plants and we would like to discuss here the former subject in relation to case number 3. In sweet potato infected with *Ceratostomella fimbriata*, Akazawa and Uritani (1955) have observed the apparent synthesis of proteinous nitrogen with concomitant increase in respiration. Tomiyama and his associates (1956) have also demonstrated the fact that in potato tissue infected by *Phytophthora infestants*, a notable increase in water-soluble protein occurs. A similar result has

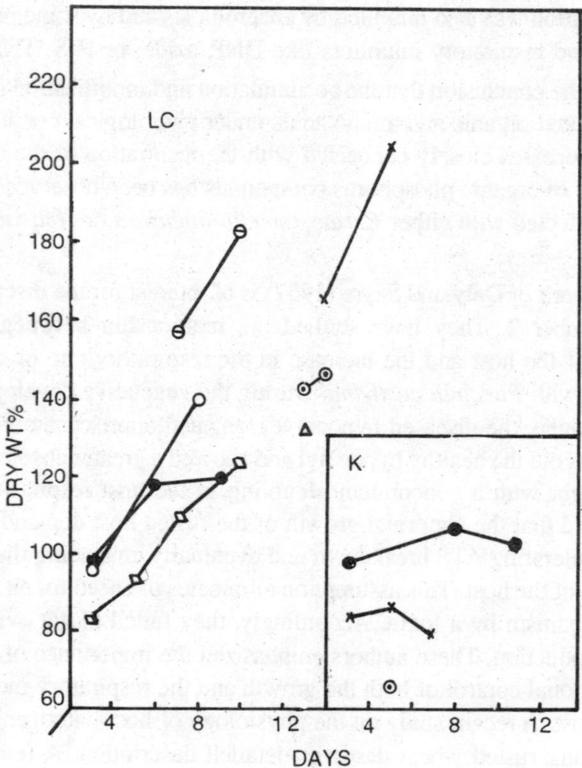

Fig. 3.3. The change in dry weight after rust infection of the first leaves of Little Club (LC) and Khapli (K) wheat species. Ordinate: dry weight of 60 leafdiscs (diameter 2.8 mm) as a percentage of the weight of discs cut frm uninfected leaves. Abscissa: days after inoculations. Each symbol represents a different experiment.

been reported in sweet potato infected with *Helicobasidium mompa*. Activation of enzymes and/or enzyme systems to be discussed later also demands energy, consequently, respiratory rate may increase.

Recent biochemical knowledge regarding the climacteric respiratory increase of fruits is pertinent to this problem. It has been generally observed that the respiratory rate of fruit tissue gradually increases in its ripening process and reaches its maximum value just before complete ripening. Analysis of nitrogenous compounds in parallel with determinations of the respiratory rate have been carried out during apple ripening.

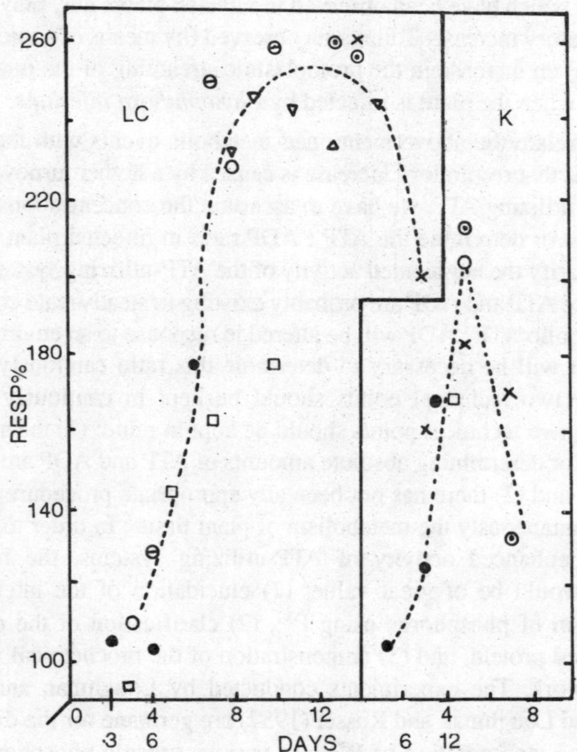

Fig. 3.4. The change in respiration after rust infection of the first leaves of Little (LC) and Khapli (K) wheat species. Ordinate: oxygen consumption per 10 mg dry weight per hour as a percentage of the value for uninfected leaves. Abscissa: days after inoculation . Symbols: as for Fig. 3.3.

Both groups have demonstrated the pronounced increase of protein content in fruit at the onset of the respiratory climacteric. Robertson's group has also noticed a slight response of tissue respiration to DNP at the climacteric stage. Their conclusion is that the climacteric respiratory increase is caused by a net synthesis of protein requiring energy and that this accelerates the turnover rate of ATP breakdown. This hypothesis is in striking contrast to that of an uncoupling one for the climateric rise of avocado proposed by Millerd et. al. (1955).

As to case number 4, all protoplasmic work demands energy. Increases in protoplasmic streaming and permeability and viscosity decreases which have been observed in infected plants may play a part in the respiratory increase. Tomiyama observed (by means of a microscopic technique) an increase in the protoplasmic streaming of the midrib cells of potato when the plant is infected by *Phytophthora infestans*.

To correlate the above mentioned metabolic events with the present hypothesis that respiratory increase is caused by a higher turnover rate of reactions utilizing ATP, we have to ascertain the concentrations of ADP and Pi and/or determine the ATP : ADP ratio in infected plant tissue as well as clarify the augmented activity of the ATP-utilizing systems of the host. Since ATP and ADP are probably existing in steady state concentrations, the ratio ATP : ADP will be altered in response to an environmental change. It will be necessary to determine this ratio cautiously and the following two technical points should be kept in cautiously and the following two technical points should be kept in mind: (1) the analytical methods for determining absolute amounts of ATP and ADP are not very sensitive, and (2) there has not been any appropriate procedure for stopping instantaneously the metabolism of plant tissue. In order to substantiate the enhanced activity of ATP-utilizing systems, the following projects would be of great value: (1) elucidation of the intermediary metabolism of phosphorus using P^{32}, (2) clarification of the nature of synthesized protein, and (3) demonstration of the biochemical nature of cellular work. The experiments conducted by Loughman and Martin (1957) and Loughman and Russel (1957) are germane for the discussion of (1). The incorporation of P^{32} into various organic phosphorus compounds in intact barley roots was investigated by using paper chromatography and a sensitive recording counter. After short periods of phosphate absorption, an intensive uptake of P^{32} into the acid soluble phosphorus fraction which contains nucleotides was observed. This process was inhibited by DNP. There were differences in the distribution of phosphorus into the various fractions depending on the concentration of P^{32} supplied, but eventually they showed the status of phosphate metabolism of the plant tissue in the steady state condition. Application of this procedure would be of great help in elucidating the phosphate metabolism of infected plant tissue in conjunction with the respiratory increase.

Demonstrating an effect on respiration due to DNP is very frequently used to show that the phosphorylative reaction is the limiting "pacemaker" in tissue respiration. Exogenous addition of DNP to tissue will cause an extra output of respiratory oxygen, depending on the velocity of ATP-utilizing reactions in the cells being studied. Akazawa and Uritani (1955) have surveyed the DNP effect on the respiration of the sweet potato infected with *Ceratostomella fimbriata* and compared the magnitude of respiratory increase with that of uninfected plants. Two to four days after inoculation, when the respiratory increase showed a maximum value, the percentage increase of respiration due to DNP addition was considerably smaller than in the healthy control tissue. At a later stage , when the respiratory increase of the infected tissue declined, the DNP effect was almost the same as that in control tissue. Other workers using wheat seedlings infected with rust have observed a decreased respiratory rate in the DNP treated tissues in comparison with a positive DNP response in the healthy tissue. In the case of the rusted wheat studied by Shaw and Samborski (1957), a decreased percentage response to DNP was found in the infected tissue, although the actual response was the same with both infected and uninfected tissues. However, it should be emphasized that a weak response of host respiration to added DNP per se does not distinguish between the two alternative mechanisms of a rapid breakdown of ATP in cells either by (a) an uncoupling phenomenon or by (b) active metabolism which we are discussing now.

C. The Inhibition of the Pasteur Effect

In the physiological study on the rusted wheat leaves, Sempio (1950) noticed the Pasteur effect abolishment which concomitantly occurred with the respiratory increase. As described before, Allen (1953) analysed Sempio's data and proposed a mechanism for the respiratory increase by means of a Pasteur effect inhibition.

Generally, the Pasteur effect is defined as the inhibition of fermentation in living organisms under aerobic conditions. It seems reasonable to extend to definition of this phenomenon to include reduction of charbohydrate breakdown under aerobic conditions with slight formation of a fermentative product like ethanol or lactic acid. Therefore, the abolition of the Pasteur effect will be considered to be a release of the suppression of aerobic carbohydrate breakdown as compared to the anaerobic one. Many hypotheses have been proposed to interpret the Pasteur effect. According to Johnson (1941) and Lynen and Königsberger (1951), the Pasteur effect would be interpreted as follows: due to the effective operation of oxidative phosphorylation more ATP is made available via the aerobic breakdown of carbohydrate than is formed under anaerobic conditions. This results in a decrease in the concentrations of ADP and Pi which influences the activity of the glyceraldehyde dehydrogenase system requiring ADP, Pi, and DNP. This enzyme is believed to be one "pace-maker" of glycolysis and a decrease in the activity of this enzyme system will subsequently decrease the rate of operation of the glycolytic pathway. The finding that coupling agents such as DNP and *p*-nitrophenol inhibit the Pasteur effect supports the hypothesized mechanism of Johnson and Lynen. Australian workers obtained favorable data that the Pasteur effect in plant tissue can be regulated by the concentrations of the phosphate acceptor system. Provided this mechanism operates *in vivo* unequivocally, the Pasteur effect will be inhibited by any metabolic event which will increase the concentration of ADP and Pi, either by the uncoupling or by the increased rate of cellular reactions requiring ATP. This is, therefore, very closely connected with the mechanism of increase in the respiratory rate.

Allen's idea is that some toxins secreted by the pathogens might cause the abolishment of the Pasteur effect by their uncoupling action, subsequently increasing the respiratory rate. Farkas and Király (1955) have proposed that the inhibition of the Pasteur effect is probably related to the respiratory increase in wheat leaves infected with rust and mildew. In this case, they assumed that it might be inhibited by the uncoupling mechanism, but were unsuccessful in attempting to isolate a phytopathogenic toxin. Similarly, Japanese workers have shown that the abolishment of the Pasteur effect in sweet potato (especially in the case of a susceptible variety) infected with *Helicobasidium mompa* is associated with a respiratory increase. American workers have obtained similar experimental evidence for an inhibition of the Pasteur effect in rust infection of safflower hypocotyl. The Canadian group has also noticed a marked reduction or perhaps absence of the Pasteur effect concomitant with the stimulated respiratory rate of rusted wheat.

In the field of plant physiology very often a ratio of anaerobic to aerobic carbon dioxide output

$$Q_{CO_2}^{N^2} : Q_{CO_2}^{O_2}$$

larger than 0.33 is considered to indicate the existence of the Pasteur effect in the organ or tissue examined. In both American and Canadian reports, the ratios of the infected rusted plant tissues fall after inoculation to below 0.3 at the final stage of the infection as compared to those for uninfected plants which do not decrease markedly from a value of 0.5 to 0.6. There was no increased anaerobic carbon dioxide formation accompanying the pathogenically increased aerobic carbon dioxide formation. From these observations, they assumed the predominant participation of the H.M.P. pathway in the respiratory pattern of infected plants and obtained supporting experimental evidences as described in the next section. The predominant participation of this pathway in the infected plant tissue would reduce the numerical figure of

$$Q^{N^2}_{CO_2} : Q^{O_2}_{CO_2}$$

and may explain the abolition of the Pasteur effect.

It has been shown recently that the Pasteur effect occurs when mitochondrial preparations exist together with a glycolytic system. The data of these workers suggest that suppression of the carbohydrate breakdown might not be regulated by the concentrations of ADP and Pi, but by some unstable system existing in a steady state form in the oxidized mitochondria. This system may be inhibiting hexokinase or phosphofructokinase. Providing this mechanism operates *in vivo*, carbohydrate breakdown could be regulated by "oxidized mitochondria" rather than by the concentration of ADP in living organisms. This hypothesis is reminiscent of the theory of Engel'hardt and Sakov (1943) on the mechanism of the Pasteur effect, which was proposed 15 years ago. According to these workers, phosphofructokinase is very sensitive to oxidizing agents generated in respiratory oxidation. It would be interesting to investigate in diseased plants the mechanism of abolition of the Pasteur effect in terms of these theories.

None of these hypotheses can thoroughly explain abolition of the Pasteur effect in diseased plants. In addition to the determination of the

$$Q^{N^2}_{CO_2} : Q^{O_2}_{CO_2}$$

ratio, more detailed analytical data are necessary before the phytopathological role of the Pasteur effect can be fully understood, e.g., the consumption of the carbon source and formation of fermentation products.

D. Activation of Enzymes Systems

Probably, the concentration of enzymes and their substrates is affecting the over-all rate of respiration. Many experiments have been published dealing with the activation or alteration of respiratory enzymes with particular reference to terminal oxidases in infected plants. Frequently, some specific substances such as polyphenols or ascorbic acid have also been found to accumulate in infected host plants. From these observations, we are inclined to assume that these anomalous metabolic events may somehow be concerned with the mechanism of respiratory increase. First we would like to discuss some general aspects of terminal oxidases in plants.

What constitutes the normal terminal respiratory oxidase in higher plants is still debated. It is generally believed, however, that the terminal electron transport system for living organisms, including plants, is mediated by the cytochrome system. Polyphenol oxidase, ascorbic acid oxidase, and some other soluble oxidases are said to function in the respiratory pattern of certain plant organs at a specific stage of their development. Mostly, such claims rest on (1) experimental results using specific inhibitors of respiratory oxidation on homogenates or tissues, or (2) failure to demonstrate cytochrome oxidase activity in the plant systems examined. Since the evaluation of these techniques has been discussed by several authors, a detailed description is omitted and basic points will be emphasized. At the present time there are two appropriate methods available for distinguishing between cytochrome oxidase and the copper enzymes (e.g., ascorbic acid oxidase and polyphenol oxidase); either by testing the photoreversibility of carbon monoxide inhibition of each enzyme, or by examining the potentiality of oxidative activity under various tensions of oxygen. In the first method, carbon monoxide inhibition of cytochrome oxidase is completely reversed by light whereas with copper enzymes it is not. (It has been occasionally reported that carbon monoxide does not inhibit ascorbic acid oxidase). In this case, transparency of tissue to light should be carefully checked. As to the second method, the affinity of cytochrome oxidase for oxygen is very high—that is, it functions normally even at low oxygen tension. On the other hand, copper enzymes have a low affinity for oxygen. Here the factor of the rate of oxygen diffusion should be kept in mind, and low activity of tissue respiration under low oxygen tension per se does not necessarily indicate the functioning of a copper enzyme. By using the above techniques, Thimann's group has clearly shown the sole function of cytochrome oxidase in the respiration of the pea internode, although ascorbic acid oxidase activity is very high in this tissue. This result is in striking contrast to the developmental study of barley roots, in which ascorbic acid oxidase has been reported to progressively replace cytochrome oxidase during the first 7 days of root development.

There have been many reports concerned with the activation of copper-containing enzymes in infected plant tissues (polyphenol oxidase and ascorbic acid oxidase), but unfortunately, none of them has been performed by carefully considering the above problems. A few examples cited hereafter will be of some value for further discussion. Király and

Farkas (1957) have examined the respiratory alteration of the wheat seedlings infected with stem rust, and from experimental results using specific substrates and inhibitors have come to the conclusion that the respiratory pattern of the plants shifts drastically to an ascorbic acid oxidase system shortly after the infection. Although this is an interesting observation, more detailed information will be necessary before a final conclusion can be reached. Uritani and Iechika (1953) have shown the accumulation of large amounts of ascorbic acid in sweet potato tissue infected by *Ceratostomella fimbriata*. Beevers (1954) has observed the oxidation of DPNH via an ascorbic acid oxidase system in cucumber; a survey of this system would be interesting in connection with the above findings.

In the case of sweet potato infected with the black rot fungus, activation of cytochrome oxidase has been confirmed in cell-free systems. However, from the experimental results of carbon monoxide inhibition and its photoreversibility in the respiration of the infected sweet potato, a part of the terminal oxidase appears likely to be mediated by a non-cytochrome system such as polyphenol oxidase. Since there is a possibility that the infected sweet potato slices are not transparent enough to light, this is not a conclusive result. Polyphenol oxidase is a widely occurring oxidase in the plant kingdom, and its activity is particularly augmented under pathological conditions. Rubin's group in Russia put forward the idea of an alteration and activation of specific respiratory enzymes from their extensive work on the physiology of host-pathogen relationship. One example is that potatoes infected with *Phytophthora infestans* show a considerable increase in the activity of polyphenol oxidase and the amount of polyphenols. They feel that this oxidase plays a significant role in the respiratory pattern of the diseased plant and in the immunity of the plant. Another example is the study of cabbage infected by *Botrytis cinerea*, in which the enhancement of peroxidase activity was observed in striking contrast to the complete impairment of other respiratory systems such as cytochrome oxidase and ascorbic acid oxidase. An interesting fact is that a similar metabolic disturbance can be induced by the *Botrytis* toxin obtained from culture in filtrates. From these observations, it was proposed that either *Botrytis* infection of cabbage or the toxin per se might switch the respiratory pattern of the host to a flavoprotein system, which will produce hydrogen peroxide by the oxidation of the substrate. The hydrogen peroxide produced by the operation of the new pathway is presumably broken down and removed by means of peroxidase. Both the role of the flavoprotein enzyme and the function of peroxidase in the terminal respiratory pattern of cabbage are totally unknown at the present time. Rubin himself does not as yet have any experimental evidence that a flavoprotein enzyme is actually activated and is functioning in the *Botrytis*-infected cabbage. This attractive hypothesis must, therefore, be proven by future study. Activation of a flavoprotein enzyme in rice plants (in the case of a susceptible variety) infected by *Piricularia oryzae* has been reported by Toyoda and Suzuki (1957).

In conclusion, the authors' opinion is that there have not been any thorough supporting data for the participation of a noncytochrome system (polyphenol oxidase, ascorbic acid oxidase, peroxidase, or flavoprotein) in the pathogenically induced respiratory increase of plant tissue. It is highly desirable to get more data on this problem. At the same time, it is important to clarify the activation of enzymes or enzyme systems other than terminal oxidases, for example, those of the TCA cycle, and the glycolytic and the H.M.P. pathways.

II. THE HEXOSE MONOPHOSPHATE PATHWAY

During the course of study on respiratory increase, experimental evidence accumulated which pointed to the operation of a new pathway, the H.M.P. pathway, in the respiratory pattern of infected plants. for instance, the following characteristic features were noticed in the respiration and carbohydrate metabolism of infected plant tissue: (1) insensitivity of tissue respiration to malonate, a competitive inhibitor of succinic dehydrogenase, thus ruling out the stimulation of operation of the TCA cycle; (2) a relatively weak response of increased respiration of NaF, the magnitude of NaF sensitive respiration being the same between infected and uninfected plant tissues, which may indicate the participation of a non-glycolytic pathway in the augmented respiration; and (3) a remarkably low value for the ratio in infected tissue,

$$Q_{CO_2}^{N^2} : Q_{CO_2}^{O_2}$$

indicating the apparent absence of a Pasteur effect. However, the magnitude of anaerobic CO_2 evolution remains approximately the same between infected and uninfected tissues as has already been discussed in C of the previous section.

Both American and Canadian workers succeeded independently in obtaining evidence for the operation of the H.M.P. pathway in rusted safflower and rusted wheat or mildewed barley. Noticeable evolution of $C^{14}O_2$ from glucose-1-C^{14} by rusted wheat leaves had been already demonstrated by the latter group. Measurement of the so-called C-6 : C-1 ratio by both groups showed the predominant participation of the H.M.P. pathway in the infected plant tissues. Here it would be helpful to describe the general sequence of the H.M.P. pathway which is diagrammatically shown in Fig. 3.5 together with the glycolytic pathway.

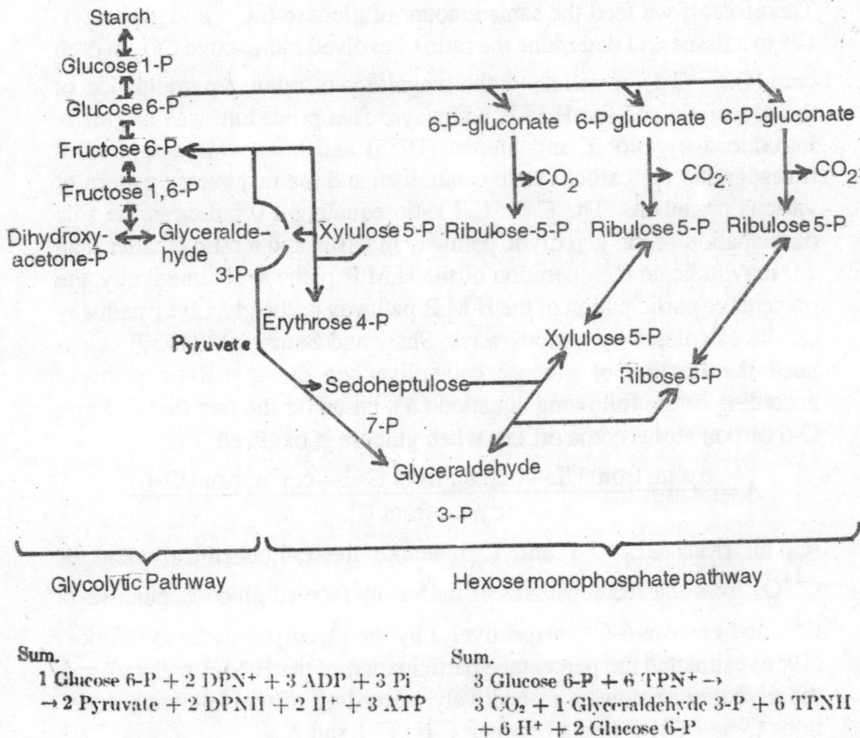

Fig. 3.5. The glycolytic and hexose-monophosphate pathways.

In the first step of the H.M.P. pathway, glucose-6-phosphate is dehydrogenated in the presence of TPN by Zwischenferment or glucose-6-phosphate dehydrogenase, and 6-phosphoglucono-γ-lactone is formed. In the following step, 6-phosphogluconate, which is formed from the phosphoglucono-lactone by the action of lactonase, is decarboxylated oxidatively also in the presence of TPN to give a pentose phosphate.

Eventually this pathway ends up by either forming phosphogleceraldehyde, which is metabolized via the glycolytic pathway, or by condensing two moles of triose phosphate to regenerate a hexose phosphate which can re-enter the H.M.P. pathway. As would be expected, if glucose is catabolized via the glycolytic pathway, CO_2 initially comes off equally from position C-3 and C-4 of the glucose molecule, and lastly from C-1 and C-6 again with equal activity. On the contrary, if glucose is oxidized via the H.M.P. pathway, the C-1 position will contribute most to the evolved CO_2 and C-6 the weakest during short term experiments. Therefore, if we feed the same amount of glucose-6-C^{14} and glucose-1-C^{14} to a tissue and determine the ratio of evolved radioactive CO_2 in each case (C-6 : C-1), it will show the magnitude of relative participation of the glycolytic and the H.M.P. pathways. This procedure was originally introduced by Bloom and Stetten (1955) and later widely applied in investigating the carbohydrate catabolism and the respiratory pattern of various organisms. The C-6 : C-1 ratio equaling 1.0 indicates the sole participation of the glycolytic pathway in tissue and a ratio smaller than 1.0 may indicate the operation of the H.M.P. pathway. Numerically, the percentage participation of the H.M.P. pathway or the glycolytic pathway can be calculated by various ways. Shaw and Samborski (1957) calculated the fraction of glucose catabolism via the glycolytic pathway according to the following equation (A), based on the fact that C-1 and C-6 carbon atoms come off last when glucose is oxidized

$$A = \frac{\text{c.p.m. from UL—(c.p.m. from C–1—c.p.m. from C–6)}}{\text{c.p.m. from UL}}$$

(c.p.m. from UL, C-1 and C-6 denote the radiochemical yield of $C^{14}O_2$ from the fed substrates of uniformly labeled glucose, glucose-1-C^{14}, and glucose-6-C^{14} respectively) by the glycolytic pathway. Korkes (1956) estimated the percentage participation of the H.M.P. pathway and the glycolytic pathway, respectively, according to the following equations (*S* and *G*), *G* being between C-6 : C-1 and A.

$$S = \frac{1 - R}{1 + 5R}, G = \frac{6R}{1 + 5R}$$

(S = fraction of CO_2 via the H.M.P. pathway, G = fraction of CO_2 via the glycolytic pathway. R = C–6 : C–1).

The data of Daly and his associates show that the C-6 : C-1 ratio of healthy safflower is around 0.6, but in the rust infected plants it decreases gradually to below 0.2 at the later stages of infection. This suggests that a striking replacement of respiratory pattern from the glycolytic pathway to the H.M.P. pathway may be evoked in the safflower hypocotyl by the rust infection. However, there is a possibility that such a change in the respiratory pattern could be due to an increase in the amount of the mycelium of the parasite, respiring through the H.M.P. pathway. Interestingly enough, DNP does not reduce the C-6 : C-1 ratio of the uninfected tissue. Also, the DNP induced respiration is NaF sensitive. Since glycolysis is selectively inhibited by NaF, the nature of the pathogenically increased respiration is substantially different from that of the DNP induced one. The experimental results of the Canadian workers are more or less the same. The c-6 : C-1 ratio of the rusted wheat dropped to around 0.1–0.3 as compared to 0.5–0.6 of the uninfected wheat leaves. As shown in Fig. 3.6, they have clearly established the positive relationship between the magnitude of respiratory increase and the percentage participation of the H.M.P. pathway in rusted wheat and have shown an increase of the C-6 : C-1 ratio in both the uninfected and the infected plant tissues due to DNP addition. Applying the above described equations, different pictures of the contribution of the glycolytic or H.M.P. pathway in the rusted safflower and wheat can be obtained, but in every case they substantiate the role of the H.M.P. pathway in the respiration of infected plants. These observations have not only explained the marked output of CO_2 by aerobic glucose breakdown, but also may give a new approach for the elucidation of carbohydrate metabolism and the respiratory pattern of infected plants.

We would now like to discuss some physiological aspects of the H.M.P. pathway in the metabolism of infected plants. It has already been observed that the pattern of carbohydrate metabolism is altered as the plant ages. That is, in the younger stages of plant growth, the glycolytic pathway is predominant, but as the plants grow older and differentiate, the H.M.P. pathway tends to replace it. Also, it has been observed that with the addition of plant growth hormones, indoleacetic acid, and 2, 4-dichlorophenoxyacetic acid, the C-6 : C-1 ratio decreases, thus indicating the hormonal control of respiratory pattern. The above described augmented activity of the H.M.P. pathway in infected plants is another case of alteration of carbohydrate metabolism in plants. As has been previously mentioned, living organisms are able to adjust their metabolic pattern to a change of environment. Presumably, mutual intervention of main and supplementary pathways in regulating metabolism is a rather common feature in living organisms. Both the glycolytic and the H.M.P. pathways appear to be concurrently taking part in carbohydrate break-

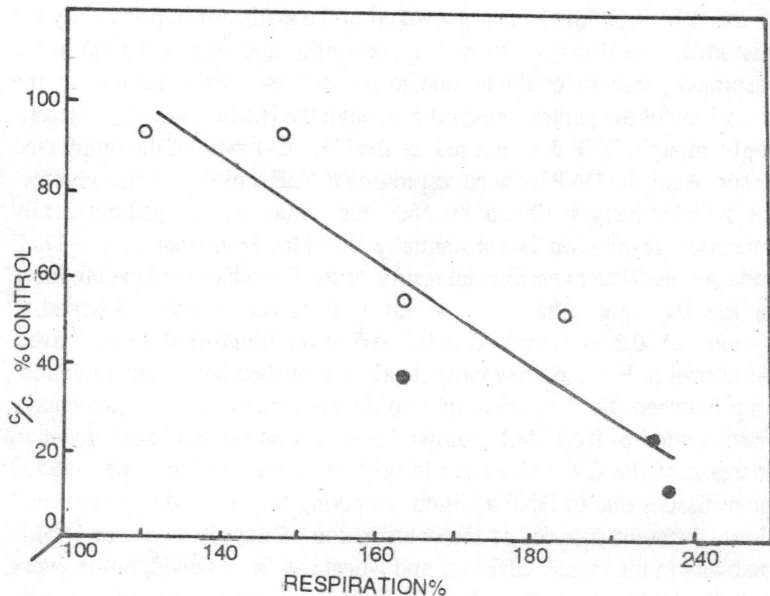

Fig. 3.6. The correlation between the C-6 : C-1 ratio and oxygen uptake for rust-infected first leaves of Little Club and Khapli. Ordinate: C-6 : C-1 ratio as a percentage of the value for uninfected controls. Abscissa: oxygen uptake as a percentage of the value for uninfected controls. Symbols: • rusted Little Club o rusted Khapli.

down and are regulating the metabolism according to the energy requirement of cells. If the requirement becomes larger, functioning of the H.M.P. pathway might increase. It is not difficult to envision vigorous energy consumption in infected plant tissues accelerating the operation of the H.M.P. pathway, though the mechanism for inducing this is not certain. We suspect that the capacity of the glycolytic pathway may not be sufficient to fulfil the total energy requirement of the host. It has been proposed that the inhibition of either phosphofructokinase or hexokinase by oxidative systems may suppress the operation of the glycolytic pathway. Furthermore, 6-phosphogluconate inhibits phosphoglucomutase. This fact may also explain the regulation of two alternative pathways. None of these possible implications can explain the reason for the more predominant participation of the H.M.P. pathway in infected plants, and in order to comprehend more fully the precise mechanism involved in the altered carbohydrate metabolism, the level of TPN (both oxidized and reduced forms), glucose-6-phosphate, and the net synthesis of the respective enzymes participating in the H.M.P. pathway in the cells should be investigated.

Thus far no experiments have been reported which deal with the interaction between the H.M.P. pathway and the respiratory chain systems, but this is also an important research subject necessary for the elucidation of the mechanism of respiratory increase. Transhydrogenase may be the most important enzyme in electron transfer systems from TPNH to oxygen. This enzyme takes part in the transfer of hydrogen between DPN and TPN according to the following equation. Evidence of whether or not the oxidation of TPNH generated via the H.M.P. pathway

$$\text{TPNH} + \text{DPN}^+ \qquad \text{DPNH} + \text{TPN}^+$$

is coupled to oxidative phosphorylation is urgently needed, and there is a suggestion that TPNH coupled to phosphorylation via the above transhydrogenase.

Aromatic compounds often accumulated in infected plants are possibly synthesized via the H.M.P. pathway. Also it has been shown that ribose-5-phosphate is synthesized by this pathway. Since ribose-5-phosphate is a constituent of ribonucleic acid, it would be of value to study the relationship between the H.M.P. pathway and the nucleic acid metabolism of virus-infected plant tissue (see Fig. 3.5).

III. BIOCHEMICAL CHANGES ACCOMPANYING THE RESPIRATORY INCREASE

As has been briefly mentioned in Section I, D, abnormal metabolites such as coumarin and polyphenols are frequently synthesized in diseased plants (Fig. 3.7). Formation of polyphenols is a common feature occurring concomitantly with the respiratory increase; and coumarins have sometimes been observed to accumulate in infected plants. Studies on the mechanism of biogenesis of these aromatic substances have made considerable progress during the last several years. This is mainly due to the advancement of research of the biosynthesis of the aromatic amino acid—phenylalanine—by the use of genetic mutants of microorganisms and radioisotopic techniques. There are still some hypothetical points in the biosynthesis of phenylalanine and all of the intermediates have not been isolated and identified; however, erythrose-4-phosphate and phosphoenolpyruvic acid have been found to be the precursors which form the aromatic rings. Participation of sedoheptulose-7-phosphate as a

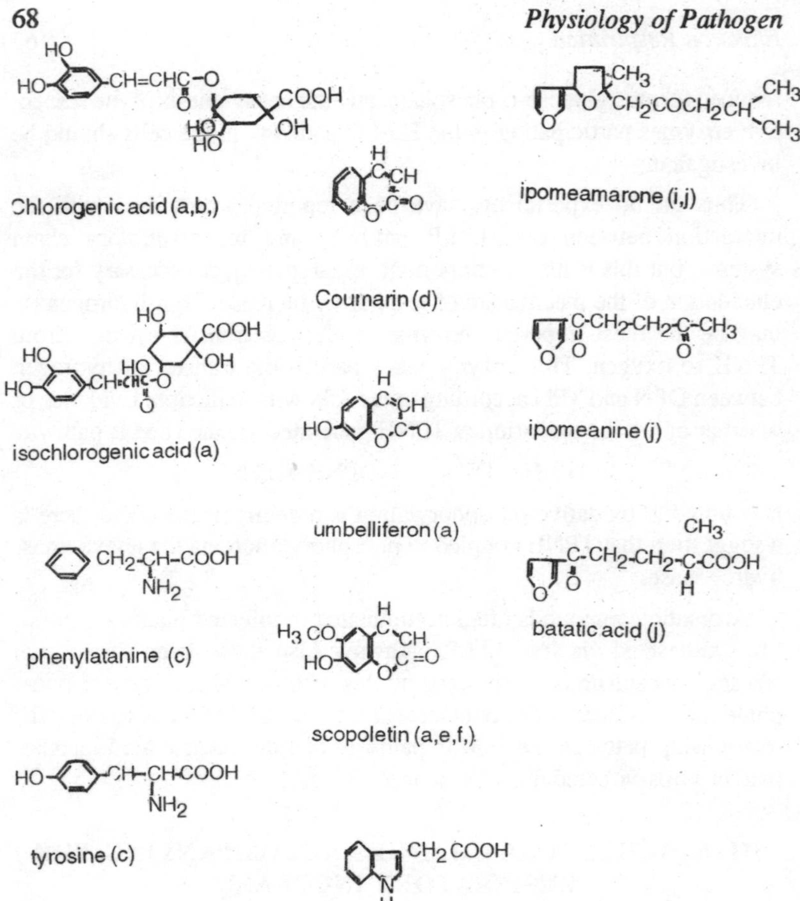

Fig. 3.7. Some metabolites accumulated in infected plants.

direct precursor of shikimic acid has been disproved; instead, it acts as a precursor of erythrose-4-phosphate. Thus, it has been shown that sedoheptulose-7-phosphate is efficiently incorporated into phenylalanine. The erythrose-4-phosphate is obviously formed from sedoheptulose-7-phosphate through the H.M.P. pathway by the action of transaldolase. If other aromatic compounds are synthesized via routes similar to that of phenylalanine (see Fig. 3.5), perhaps the important function of the H.M.P. pathway is in the biosynthesis of various types of aromatic compounds. A predominant participation of the H.M.P. pathway in infected plants may thus have a close connection with the biosynthesis of polyphenols and coumarin derivatives.

	Host	Pathogen
a	Sweet potato	*Ceratostomella fimbriata*
b	Potato	*Helminthosporium carbonum*
c	Wheat	*Puccinia graminis*
d	Rice plant	*Piricularia oryzae*
e	Tobacco	Tomato spotted wilt virus
f	Potato	*Phytophthora infestans*
g	Wheat	*Puccinia graminis*
h	Safflower	*Puccinia carthami*
i	Sweet potato	*Ceratostomella fimbriata*
j	Sweet potato	*Ceratostomella fimbriata*

The effect of rust infection on the amino acid metabolism of wheat leaves has been studied by two groups. Shaw and Colotelo have shown that in the susceptible species of wheat infected with stem rust several amino acids, including aromatic amino acids, have increased notably as compared to the levels in the uninfected plant. The amino acid composition of protein hydrolyzsates of the rusted wheat has also greatly changed, and they suggested that an alteration in the enzyme systems concerned with the metabolism of those compounds may occur under pathological conditions.

There are several reports indicating an increase in the auxin concentration of the infected plant, though it is very low indeed in a healthy one. All of them have noticed hypertrophy or cell elongation of the host when infected and have discussed this phenomenon with respect to the influence of auxin. Shaw's group analyzed the amount of indoleacetic acid in a quantitative manner and simultaneously looked into the enzymatic oxidation of radioactive indoleacetic acid ($C^{14}OOH$) by analyzing the release of $C^{14}O_2$. The concentration of indoleacetic acid in the infected leaves of the susceptible species (Little Club) of wheat decreased initially after infection but 10 days after infection was ten to twenty times higher than that of the uninfected leaves. A reciprocal relationship was observed in the activity of indoleacetic acid oxidase. However, no such increase of indoleacetic acid was seen in the case of infected wheat leaves of the resistant species (Khapli). This is, perhaps, due to a constant higher activity of indoleacetic acid oxidase in the infected plant of the resistant species. The American group analyzed the auxin content of safflower hypocotyl infected with *Puccinia carthami* during the course of infection and obtained results similar to those of the above authors. Shaw and his co-workers (1958) have observed that indoleacetic acid lowers the C-6 : C-1 ratio. From this fact together with the lowered C-6 : C-1 ratio in rusted and mildewed leaves, it seems probable that the auxin is conceivably somehow controlling the respiratory metabolism of the host tissue. However, firm conclusions should not be drawn from the following observations: (1) lower C-6 : C-1 due to indoleacetic acid addition is attributable to the lower recovery of C-6 whereas it is due to the higher recovery of C-1 in the rusted wheat leaves; (2) also, as has been described above, there was an apparent decrease of indoleacetic acid content in the resistant species of wheat leaves, irrespective of the lower C-6 : C-1 ratio.

Three compounds have been isolated from the sweet potato infected by the black rot fungus (i.e., ipomeamarone, ipomeanin, and batatic acid) and their chemical structures (see Fig. 3.7) proven. As the chemical structure of ipomeamarone shows, this is a kind of sesquiterpenoid with a molecular formula of a reduced type. We feel that these compounds may be synthesized by an unknown alternative pathway of the host metabolism. Metabolism of C_2 compounds, which are normally oxidized by the TCA cycle or sometimes utilized to synthesize fatty acid or phospholipid, may be bypassed to form isoprenoid units and eventually lead to a synthesis of ipomeamarone and its derivatives.

IV. ALTERATION OF THE RESPIRATION IN RELATION TO DEFENSE MECHANISMS OF THE HOST

Sufficient data are not available to formulate a straightforward hypothesis on the relationship between the altered respiratory pattern in infected plant tissue and defense reactions, but some speculation here may guide future study. Conceivably, metabolic patterns evolved in plants allowing them to adjust to a changing environment. Probably, defense devices of plants have been improved from a very primitive to a highly advanced type. As stressed repeatedly, the increase in respiratory rate is a comprehensive phenomenon observed in many plants exposed to varied stimuli ranging from a simple mechanical treatment to a complex pathogenic infection. Therefore, it is important to realize that respiration of the host tissue increases in response to an environmental change. This applies also to the defense mechanisms of plant tissue against pathogens as well as other injurious agents.

From a large amount of experimental data, it appears that respiratory increase and accompanying metabolism are more or less parallel to the magnitude of the pathogenic penetration of the parasite in the host. The more severe the pathogenic infection becomes, the more intensive is the metabolic activity. Thus, an augmented respiratory rate is an expression of the host response to the penetrating pathogen. However, this does not mean that the magnitude of the respiratory increase is a measure of the resistance activity of host tissue. In the infection of rust resistant wheat, the data of Shaw's group have shown a rapid respiratory increase followed by a sharp drop of respiratory rate as compared to the respiratory rate in the susceptible species. The magnitude of increased respiration was not, however, greater than that observed in the susceptible species (Fig. 3.4). Such features have been observed by several workers. In the case of the potato-*Phytophthora* complex studied by Tomiyama and his coworkers (1956), the change in the respiratory rate of the resistant variety is more pronounced than that of the susceptible variety both in the sharp rise and in the total amount of oxygen uptake. Observations of Uritani and Akazawa (1955b) on the black rot disease of sweet potato indicate a more rapid increase and a greater magnitude of the respiratory rate in the resistant than in the susceptible varieties (Table 3.2). From these observations, it appears that the initial sharp rise in the respiratory rate of infected plant may be connected with the defense mechanism of the plants.

In addition, several people have found that they can efficiently diminish the resistant activity of the host by inhibiting the phosphorylation reaction. In other words, the pathogen is able to penetrate the host more readily following a treatment of this kind. For instance, if a sweet potato infected with *Ceratostomella fimbriata* is left under conditions of high humidity and low oxygen tension, no marked enhancement of metabolism of the host is observed; rather, the fermentation process predominates in the host tissue. Under these conditions, the pathogen spreads over the surface of the sweet potato and penetrates readily. Several experimental results have shown that treatment of potato tubers with ethanol facilitates greatly the penetration of *Phytophthora infestans*. According to Tomiyama's experiment, processes such as respiratory increase, syntheses of water-soluble protein and polyphenols are suppressed considerably in the presence of ethanol; accompanying this is a decrease in the resistant activity of the host. Similar observation has been reported for rice plants infected with *Piricularia oryzae*. Walker and his co-workers have shown that tomato plants treated with some respiratory inhibitors, including the uncoupling agent DNP, lose their resistant activity to *Fusarium oxysporum* f. *lycopersici*. They have suggested that some very labile substance, which is closely connected to the phosphate metabolism of the host, might be an important factor in the resistant activity of tomato. According to Tomiyama's cytological study, midrib cells of potato have shown the following sequential microscopical features when infected with *Phytophthora infestans*. In most cases, immediately after the active protoplasmic streaming, hypersensitive death of cells follows. However, if the tissue is treated with DNP, the

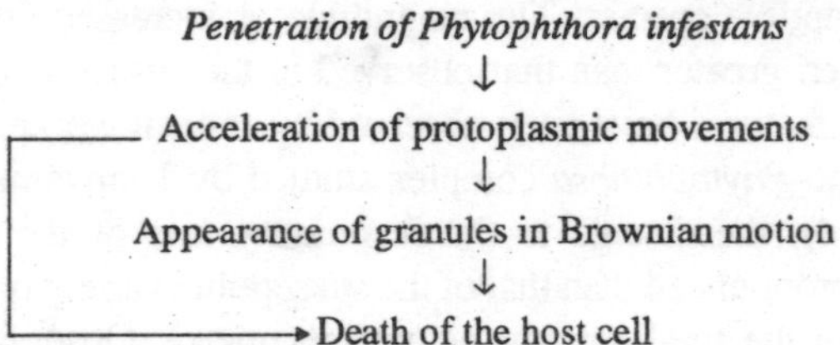

death of the host is remarkably delayed and it becomes susceptible to the parasite. As these several experimental results show, there definitely exist defense mechanisms of the host which demand respiratory energy that is possibly linked to the phosphorylation system.

Accompanying the respiratory increase, many compounds including polyphenols and coumarins are synthesized and some of them may be functioning as antipathogenic principles of the host. Kuc and his associates (1956) have studied the accumulation of polyphenols in potato tubers infected by *Helminthosporium carbonum* and suggested the role of the substances in natural immunity of plants. In the case of the black rot of sweet potato, 1–5 mg polyphenols per gram fresh weight tissues accumulated. However, with these concentrations, there is little inhibition of the black rot fungus. Apparently there is no positive relationship between the invasion of the pathogen and the amount of polyphenols synthesized (Table 3.2) in the host. Studies of the antipathogenic nature of polyphenols in infected plants and the role of polyphenol oxidase in relation to the physiological action of polyphenols are (from a biochemical viewpoint) most pertinent in the field of phytopathology. Coumarin substances were isolated from the diseased plants and some of them exhibited fungitoxin action. Ipomeamaron, found in the injured part of the diseased sweet potato is definitely an anti-parasitic substance (Table 3.2) anJ the amounts are roughly parallel to the degree of resistance of the plant.

When a resistant variety of plant is attacked by a pathogen, instant necrosis (denaturation, coagulation of proteins, and browning) precedes the penetration of the pathogen into healthy tissue. This phenomenon is called hypersensitivity. Probably, hypersensitivity plays an important role in the defense mechanism of the host against the pathogen by producing antipathogenic agents and forming protecting barriers. Müller (1956) isolated some substances from the inner epidermis of pea pods infected by *Sclerotinia fructicola* and *Phytophthora infestans* which have an inhibitory action against the pathogen and also induce necrosis of the host. Rubin feels that the activation of polyphenol oxidase in potato tissue infected with *Phytophthora infestans* is the crucial factor in inducing necrosis and the excellent work of the German researchers is considered to coincide with this opinion. These workers found that by treating the potato tuber with some respiratory inhibitors, particularly by copper chelating agents, the host became susceptible to *Phytophthora infestans*. Japanese workers have emphasized the fact that hypersensitivity accompanying the respiratory increase is an important factor in the resistant action of the host. Hypersensitivity is the result of the interaction between host and pathogen. Although the mechanism of the phenomenon has not as yet been clarified, we might postulate the following: when attacked by the pathogen, the metabolism of the host cell is activated. Some enzymes and/or metabolites produced by the pathogen damage the protoplasmic function of the cell. When the cell is injured some substances produced by it stimulate necrosis, thus leading to hypersensitivity.

Yet, the fact that respiratory increase does not always accompany hypersensitivity should be emphasized. For instance, when a nonresistant sweet potato variety is attacked by *Ceratostomella fimbriata*, the above described features of metabolic alteration are observed and ipomeamaron is synthesized in the injured area. The pathogen continues to penetrate gradually into the host tissue, yet profound hypersensitivity is not observed. We may postulate the production of hypersensitivity stimulators by host cells. In the case of sweet potato, nonresistant to the black rot fungus, the potency of these two factors might not be strong enough to cause necrosis preceding penetration of the pathogen. Those factors are hypothetical ones and Müller's phytoalexins, Rubin's activated polyphenolase-polyphenol system, and ipomeamarone and other abnormal metabolites in the infected sweet potato might be considered to be examples of hypersensitive stimulators. Study on the function of necrosin discovered in the inflammation of infected animal tissue, would be of value in elucidating the function of this proposed factor.

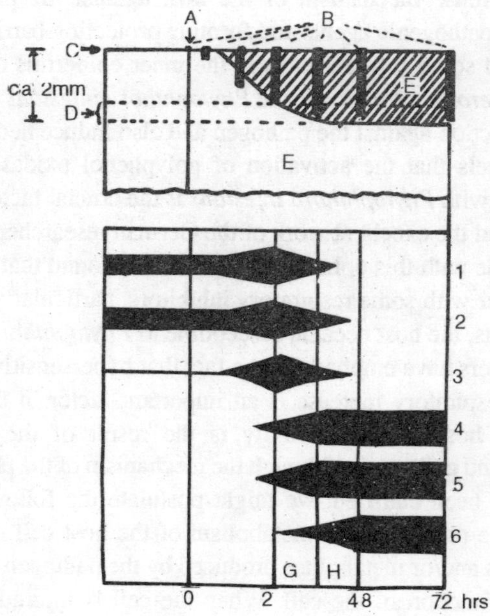

Fig. 3.8. The relationship between change of respiration in the infected sweet potato and its resistance against *Ceratostomella fimbriata*. The upper part of the diagram shows development of fungus penetration (B) concomitant with the progress of injured tissue (E) and the gradual decrease of healthy tissue (F). At 48 hr after inoculation (A) of the fungus on the surface (C) of sweet potato slice, the infection is apparently prevented by the resistance action of the host and further penetration is not observed. The lower part shows the quantitative change of some biochemical events (1–6) observed in (D) (the tissue 1.75 mm inside from the surface (C) as a consequence of the pathogenic infection. (1) Change of over-all ATP-utilizing system. (2) Change in the respiratory rate. (3) Change of the amount of coumarins and polyphenols. (4) Inactivation and denaturation of protein. (5) Synthesis of ipomeamarone. (6) Browning. Up to 24 hr. after the inoculation, the (D) part is healthy and not infected, but afterwards this part is gradually injured until about 48 hr., and is playing a role in the resistance during 24 to 48 hr. Presumably this stage is divided into two parts (G) and (H). At (G) stage, the resistance might be attributable to some antibiotic factors such as coumarins (3), polyphenols (3), and ipomeamarone (5). At (H) stage the resistance might be mainly due to the barrier formation in which events of (4) and (6) are involved.

Browning is a characteristic feature of hypersensitivity in the host, but the exact mechanism by which this biochemical reaction takes place is not known. Histological observations of necrosis in the injured tissue of infected plants lead us to the following assumption: first, polyphenolic compounds are oxidized by polyphenol oxidase which may exist in a latent state in the intact plant tissue and be activated on exposure to pathogenic infection. Second, the oxidized polyphenols, now quinones, are condensed to form polyquinoid structure or sometimes react with amino acids or proteins to form melanin substances. The net effect of these reactions may constitute a defense mechanism of the host by forming a barrier.

In conclusion, respiratory increase of host tissue appears to be a prerequisite for the induction of the resistant action of the plant. Some other biochemical events occurring concurrently with the respiratory alteration may amplify the resistance of the host through their anti-pathogenic action. And finally, hypersensitive reactions of the host preceding the pathogenic infection are necessary to establish the resistance action. Those over-all relationships are diagrammatically exemplified in Fig. 3.8, in which the change of respiration in the infected sweet potato and its resistance against *Ceratostomella fimbriata* are shown.

4

EFFECT ON GROWTH

One of the most striking characteristics of higher plants and animals is the extraordinary way in which all of their functional parts fall into a coherent and flexible, but definitely limited pattern. In the normal organism, morphogenetic laws are strictly obeyed and processes concerned in metabolism, growth, cellular differentiation, and organogenesis are precisely regulated. These processes start and stop in harmony to give the organism certain proportions that remain constant from one generation to the next.

The normal plant may be looked upon as a precisely balanced complex of interrelated reactions that are in a state of dynamic equilibrium. Any deviation in this complex of reactions will tend to disrupt the balanced system and may lead to atypical, abnormal, or pathological growth. It is not surprising, therefore, that growth patterns of plants may be readily modified by external environmental conditions such as heat, cold, light, and humidity, by nutritional deficiencies and excesses, by radiation, by changes in genetic constitution, as well as by pathogenic organisms of the most diverse types. The borderline between the normal and the abnormal in plants is, therefore, often quite indistinct and may represent merely quantitative changes which are evidenced either by a harmonious but somewhat exaggerated manifestation of the normal developmental processes or, as is more commonly found to be true, by an arrest or inhibition of the processes concerned with growth and development. At other times, orderly qualitative changes result which in their most interesting form reveal growth and developmental potentialities in a plant far beyond any realized in the past history of the species. In the most extreme instances, plant cells undergo far-reaching qualitative changes and may, as a result, become permanently modified into new cell types in which continued unregulated and unorganized growth, rather than differentiation, characterizes the behavior of the affected cell.

Of particular concern to the present discussion is the influence that parasites of many different kinds exert on the growth and development of a parasitized plant.

Gäumann (1954) has indicated that organisms are pathogenic only if they are toxigenic. Although this sweeping generalization probably contains a large element of truth, little is actually known yet about the physiological and chemical mechanisms underlying disease. It is nevertheless true that the growth of systemically diseased plants is generally inhibited. Even in those instances in which the disease is very mild and symptoms are largely masked (latent virus infections of potato, masked strains of tobacco mosaic virus), a statistically significant inhibition of growth of the host plant is found to occur. More often, pronounced stunting is a characteristic manifestation of disease in plants. This may result either from systemic infection, or from the diffusion of specific chemical compounds elaborated by a pathogen, which itself remains localized in the host. These chemical substances fall essentially into two categories: (1) enzymes; (2) soluble metabolites, many of which are capable of reproducing perfectly the toxic manifestations of disease and some of which have been characterized chemically.

Of far greater biological interest than the substances which act in such an unsophisticated way to damage, kill, or otherwise inhibit growth of the cells of the host, are those substances that stimulate cells to excessive growth.

The regulation of growth, differentiation, and organ formation in higher plants appears to result, among other things, from a very precisely balanced series of growth-regulating substances or hormones, on the one hand, and from inhibitory systems or other compensatory mechanisms that control the synthesis and response of plant cells to such biologically active substances, on the other. Certain metabolites elaborated by pathogens appear to be quite similar or identical in their physiological action to growth-regulating hormones found to occur naturally in a plant. When these are produced in excess by the pathogen during the course of infection, growth responses occur—in a host—that simulate those found following the artificial application of excessive amounts of such growth-promoting substances. At other times, the chemical stimuli produced by the pathogen are quite different from any thus far recognized in the normal plant, and, being different, tend to divert the normal growth pattern of the plant into new and unusual directions. Such morphogenetic stimuli appear to be of many different types. They are commonly highly specific in their action and regularly induce unusual growth responses in an affected plant. It must be recognized, however, that whatever form the growth abnormality takes, the potentialities for this form must have been present in the cells of the host. These growth potentialities are simply called into activity by the inciting stimulus. Thus, the morphogenetic stimulus produced by the pathogen as well as the protoplasmic substrate of the host upon which the stimulus acts are critical in determining the type of growth that results.

Finally, the metabolite elaborated by the pathogen may not itself be a hormone or a morphogenetic stimulus of the type referred to above which directly influences the growth pattern of the host. It may instead affect specifically the regulatory mechanism of the host cells by eliminating either temporarily or permanently those cellular systems that are concerned with the regulation of growth or, alternatively, it may activate within the affected cell growth-substance-synthesizing systems whose products are concerned specifically with growth and developmental processes. In these instances, it is the affected cell itself that elaborates growth-promoting substances—in greater than regulatory amounts—in response to a specific stimulus transmitted by a pathogen.

I. HARMONIOUS CHANGES INVOLVING EXAGGERATED GROWTH RESPONSES

A. Generalized Stimulation

1. *Bakanae Effect*. Intensified manifestations of normal developmental potentialities have been recorded in a number of plant species following infection with a variety of fungi, many of which are obligate parasites. The common houseleek, *Sempervivum hirtum*, grows as a rosette. The leaves of this plant are broadly ovate in form and are about twice as long as they are broad. Following infection with the rust *Endophyllum sempervivi* the appearance of the leaves is altered. They grow strongly in length and may be seven times as long as they are broad. As a result, the leaves assume a linear shape. Infected leaves stand erect and are much paler in color than are their normal counterparts.

Elongation of the internodes with a resulting significant increase in size appears to be a characteristic response of certain plants to infection by specific pathogenic fungi. Following infection of one of the spurges, *Euphorbia cyparissias*, by the aecial sage of the rust *Uromyces pisi*, the stem of the host elongates greatly. The distances between successive leaves in healthy *Euphorbia* plants are about 0.5 mm while those of infected plants are 2–3 mm. The foliage leaves of the normal plant are thin, flexible, and about twelve times as long as they are broad. In diseased specimens, the leaves are thick, brittle, and only about two to three times as long as they are broad. Infected plants have an etiolated appearance. Similar responses have been observed in *Vaccinium vitisidaea* infected with the teleutospore stage of *Melampsora goeppertiana*, in sugar cane with the downy mildew *Sclerospora sacchari*, and in *Bromus erectus* with the smut fungus *Ustilago hypodytes*. Alterations produced on the shoots of the periwinkles *Vinca herbacea, Vinca major* and *Vinca minor* by the uredospore stage of the rust *Puccinia vincae* and on the shoots of *Cirsium arvense* by the teleutospore stage of *Puccinia suaveolens* are also very similar to those described above for *Euphorbia*. The stems of the infected plants become much elongated, while the leaves are shorter, broader, yellow in color, and brittle when compared with the normal. Frequently, the shoots blossom prematurely and the flowers are more or less abortive. When, for example, *Primula clusiana* and *P. minima* are infected with *Uromyces primulae integrifoleae*, not only do the rosette leaves elongate but the flowers open in the autumn of the same year rather than in the following spring as happens in the normal uninfected specimens.

Pilet (1953) studied the auxin content of healthy leaves of *Sempervivum* sp. infected with *Endophyllum sempervivi* as well as of healthy and diseased leaves of *Euphorbia cyparissias* infected with *Uromyces pisi* in an attempt to account—on a physiological level—for elongation of the stem as a result of infection. Parasitized leaves of both plant species revealed a much higher auxin content than normal. On the basis of these observations, Pilet (1953) suggested (1) that the causal fungus elaborates auxins, (2) that the increased auxin level found in the host is produced by the host in response to the parasite, or (3) that the parasite produces something which activates precursors of auxins in the host cell and which in turn are converted into auxin. It has not yet been shown, however, that the application of auxin to normal plants results in an elongation of leaves and internodes comparable to that obtained after infection by specific rust fungi. Brian (1957) has, therefore, suggested that the increased auxin found in the leaves and shoots of such plants may be coincidental rather than causal since quite different metabolic products of fungi, the gibberellins, have been shown to be responsible for stem elongation in the bakanae disease of rice as well as in many other plant species when these substances are applied artificially.

The bakanae or foolish seedling, disease of rice is the best understood of this type of disease. It has been studied by Japanese pathologists for many years. A comprehensive review of this disease has recently been presented by Stowe and Yamaki (1957). The bakanae disease, which is caused by *Fusarium moniliforme* (imperfect stage) or *Gibberella fujikuroi* (perfect stage), is widely distributed and found in most rice-growing-regions of the world. *Fusarium moniliforme* has a large host range, is soil-borne, and attacks the roots and basal portions of the stem not only of rice plants but of maize, cotton, sugar cane, and other plant species as well. In the case of rice, a browning of the tissue usually occurs at the site of infection, the leaves tend to yellow and curl inward and, in many instances, growth of affected seedlings is arrested. However, some of the affected plants grow more rapidly than do healthy ones and are conspicuous in the field because of their height and etiolated appearance. The dry weight of the elongated seedlings was found to be significantly greater than that of the healthy plants. This overgrowth effect—occurring naturally—has been reported only in rice, although artificial inoculation has resulted in increased growth in maize, barley, sugar cane, sorghum, millet, wheat, and oats.

Kurosawa (1926) showed that the overgrowth or bakanae effect could be reproduced in rice seedlings by treating them with cell-free culture filtrates on which the causal fungus had grown. These filtrates reproduced all of the characteristic manifestations of the bakanae disease such as lengthening of the internodes and leaves, chlorosis (except under conditions of nutrient excess), and reduced tillering. This pioneer work of Kurosawa was followed by the isolation and chemical characterization of several biologically active substances. Two powerful growth-promoting substances, which were given the trivial names of gibberellin A and B, were isolated in crystalline form by Yabuta and Sumiki (1938), and Yabuta and Hayasi (1940). It was found later that gibberellin A is converted to B by warming in dilute acid at 50°–70°C. Boiling under acid conditions converted both substances into a biologically inactive compound named gibberic acid. A third biologically active substance, gibberellin C, was isolated by Yabuta et. al. (1941) from acid-treated gibberellin A. The nature of the gibberellin skeleton was established by Yabuta et. al. (1941). It was demonstrated in these studies that gibberene obtained by selenium dehydrogenation of gibberellin A, B, or gibberic acid is a fluorene derivative.

Another quite distinct but chemically related, biologically active compound was isolated by American workers headed by F.H. Stodola and by

British investigators under P.W. Brian. This substance was named gibberellin X by the Americans and gibberellic acid by the British group. The latter name is now commonly used for this compound. Although gibberellic acid, whose empirical formula is $C_{19}H_{22}O_6$, resembles gibberellin A in biological activity and degradation products, it differs from this compound in infrared spectra, optical rotation, in certain derivatives, as well as in the empirical formula. The basic-skeleton of gibberellic acid, like gibberellin A, was found to be gibberene, which Mulholland and Ward (1954) identified as 1, 7-dimethylfluorene. This led Cross et. al. (1956) to suggest the tentative structure for biggerellic acid shown in Fig. 4.1. It is

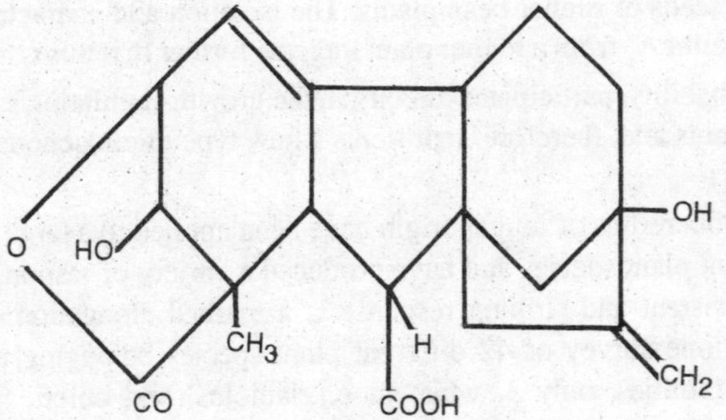

Fig. 4.1. Gibberellic acid.

reasonably certain that the formula shown in Fig. 4.1 is essentially correct, although the exact points of attachment of the lactone grouping to the cyclohexenol ring of gibberellic acid are not known as yet.

Recently, Takahashi et. al. (1955) re-examined gibberellin A and found this compound to be essentially homogeneous by countercurrent distribution, paper and partition chromatography. Esterification followed by chromatography on alumina led to the isolation of three methyl esters which were named gibberellin A_1, A_2, A_3. Stodola et. al. (1957) believe that their gibberellin A is identical with A_1, Gibberellic acid was shown to be identical with A_2. Gibberellin A_2 has been isolated only in Japan but Stodola (1956) reports its corrected empirical formula as $C_{19}H_{26}O_6$. Thus, the existence of three distinct gibberellins appears to be established.

Evidence is now available suggesting that the gibberellins elaborated by the fungus *Gibberella fujikuroi* correspond closely (in their biological activity) to growth-regulating compounds found to occur naturally in higher plants. West and Phinney (1957) have isolated—but have not yet identified chemically—an ether-extractable substance from wild cucumber seeds which, like the gibberellins of fungus origin, promotes the active growth of dwarf mutants of maize. Since these genetic dwarfs appear to be due to single-gene defects, they may be explained as resulting from blocks in the biosynthetic pathway leading to the formation of a natural gibberellin-like substance. The gibberellins appear to substitute in the plant for the missing product of that reaction. Very recently MacMillan and Suter (1958) have obtained high yields of gibberellin A_1 from the seeds of runner bean plants. The isolation and characterization of gibberellin A_1 from a higher plant indicate further that this compound, in all probability, participates directly in the growth-regulating system of higher plants and, therefore, represents a new type of endgenous growth regulator.

The gibberellins of fungal origin have been applied to a large number of different plant species and have produced a variety of responses. The most consistent and striking response is a marked elongation of plant stems. In one survey of 42 different plant species belonging to many different families, only 3, white pine, gladiolus, and onion, failed to respond with stem elongation. In many instances, not only do the stems grow longer but also they are proportionately thicker. Gibberellin-treated oak and maple trees are not only taller but have a diameter more than twice as great as that of untreated control specimens. Brian and Hemming (1955) have reported remarkable stimulation of dwarf varieties of peas, broad beans, and French bean by gibberellic acid. The treated dwarfs and untreated normal plants grew equally well.

In the bakanae disease of rice as well as in certain related diseases described earlier in this discussion, infected plants that reached maturity flowered earlier than did normal plants of the same variety. With the availability of pure gibberellic acid for experimental purposes, Lang (1956) found that a biennial variety of *Hyoscyamus* treated with that substance bolted and flowered in the first year. In a later study, Lang (1957) found that in annual varieties of *Hyoscyamus, Silene*, *Samolus* the long-day requirement for flowering could also be replaced by gibberellic acid. Wittwer et. al. (1957) found that gibberellic acid-treated lettuce colted and flowered under photoperiodically adverse conditions. These same workers reported the gibberellins to be 500 times more effective than indoleacetic acid in inducing parthenocarpy in tomatoes.

Reduced fruit yields are characteristic of the bakanae disease. Hayashi et. al. (1953) found that the gibberellins reduced rice grain production by 32% although the yield of straw was increased by 14%.

Stowe and Yamaki (1957) suggest that the one property common to the gibberellins is their capacity to remove certain limitations on cell enlargement, while incidentally releasing other responses. Despite this similarity to the auxins, the gibberellins are not auxins but appear to depend for their activity on the presence of auxins. The biological activity of these substances has been attributed to the neutralization of a growth-inhibitory system which normally limits growth.

Certain other plant diseases give indication that infection interferes with the synthesis or utilization of gibberellin-like substances. Carrot plants infected with the tomato big-bud virus of California, as well as with a related virus found to occur naturally in carrots in the State of Washington, were shown by Kunkel (1951) to bolt and flower prematurely. These findings suggest that virus infection may increase the synthesis of gibberellin-like substances by the plant. On the other hand, in the rosette diseases, as exemplified by peach and peanut rosette, infection may interfere with the normal synthesis or utilization of these biologically active substance by the plant. The stunting effects, but not other symptoms associated with aster yellows, corn stunt, and wound tumor diseases, have in fact been reversed with gibberellic acid.

B. Growth Changes in Leaves and Petioles

1. *Growth Movements*. Nastic responses refer to changes in position of a bilaterally symmetrical organ as a result of differential growth. Epinasty of leaf petioles has been found by Wellman (1941) to be a characteristic early symptom of the *Fusarium* wilt disease of tomato plants. Hunger (1901) and Smith (1920) found it to be associated with Granville wilt, while Grieve (1941) recorded this condition as a primary symptom in roses affected with the rose wilt virus. Pronounced epinasty has also been produced experimentally by Locke et. al. (1938) following inoculation of tomato plants with a highly virulent strain of the crown gall bacterium. Hyponasty, on the other hand, has been observed in plants infected with *Erwinia phytophthora*.

Grieve investigated the question of epinasty of the leaf petiole as it occurs in the Granville wilt disease (1940). In this instance, the growth response was found to be an irreversible one and the invasion of one lateral trace by the bacteria was sufficient to induce it. Grieve reported that a growth substance which appeared to be indoleacetic acid was elaborated by the bacteria in culture and that this substance induced a characteristic epinastic response when applied to tomato plant. However, virulent and avirulent cultures of the bacterium produced approximately equal amounts of the growth-promoting substance. Furthermore, no significant difference between the growth-substance content of comparable control and infected stem segments could be detected. Studies of the growth-substance distribution in upper and lower halves of reflexing petioles, on the other hand, showed a significantly greater concentration in the upper halves. In normal petioles, the concentration of growth substance was found to be greatest in the lower halves of the petioles. Grieve (1939) reached no definite conclusion as to how the redistribution of growth substance in the basal part of the petiole is accomplished. He pointed out, however that the balance of the normal growth-controlling mechanism at the base of the petiole is very delicate, as evidenced by the fact that ethylene, in one part in ten million parts of ambient air, as well as very minute amounts of growth substance from bacterial cultures will disturb it. Grieve considers that even a small stimulus from the invading organisms can initiate a chain of reactions that leads to a redistribution of growth substance with a resulting nastic response.

More recently, Dimond and Waggoner (1953) made a detailed investigation of the cause of epinastic symptoms in *Fusarium* wilt of tomato plants. From these studies it was concluded that ethylene is responsible for the characteristic growth response. This conclusion was based on the finding that ethylene is produced in culture by the causal agent of this disease, *Fusarium oxysporum* f. *lycopersici*. Ethylene production by infected tomato plants was demonstrated, moreover, by confining such plants with healthy indicator plants. Under these conditions, epinastic response developed to a greater degree than when the indicator plants were confined with healthy tomato plants. In these studies, ethyl alcohol, which is capable of causing epinasty in tomato and which is produced by a causal fungus in culture as well as in the infected host, was also considered a possible cause of the growth response. This compound was ruled out, however, because the amounts produced by the fungus in the hosts were insufficient to account for the observed epinastic responses.

Another characteristic response involving growth movements—which is associated with certain disease conditions is concerned with the upright growth habit of leaves. This is one of the most characteristic symptoms of the aster yellows disease in many different plant species. Leaves on a normal aster plant take approximately a horizontal position in relation to the main axis of the stem. In virus-infected plants, on the other hand, the petioles elongate and the leaves assume an extremely upright habit of growth which approximately parallels the main axis of the plant. It is this upright habit of growth that suggested the name "rabbit ears" for lettuce plants infected with the aster yellows virus.

Kunkel (1954) has shown that twigs of peach trees infected with peach yellow virus characteristically assume an upright growth habit. In this instance, the plants can be cured of the disease by thermal treatment. After cure, the new growth of such twigs again takes a normal position with respect to the main stem. This change necessitates a change in the direction of growth of the terminal bud.

2. *Curling and Distortion*. The curling and distortion of leaves are characteristic of many diseases. Needless to say, such well known maladies as curly top of sugar beet, peach leaf curl, and leaf roll of potatoes have been assigned their trivial names because curling of leaves is the most conspicuous symptom of the disease. The most extreme form of leaf rolling is found in the so-called scroll galls produced by certain insects. Under the influence of specific insects, the leaves curl lengthwise to form tightly rolled scrolls in which the insects live.

Peach leaf curl caused by *Taphrina deformans* is perhaps the most studied of this type of disease. The causal fungus commonly infects very young leaves which soon become either quite red or paler in color than are the normal leaves. Such infected leaves soon become curled and puckered, increase greatly in thickness, and have a firm consistency. The host cells which are in contact with the invading fungus are stimulated to abnormal activity. This may involve isolated regions of a leaf or in extreme instances most of the leaf. Such cells increase in size and number and produce marked changes in the form and structure of the leaf. Loss of chlorophyll is almost complete in the stimulated cells. The increase in size of the cells on either side of the midrib results in a puckering of the leaf. At the same time, the leaves tend to curl and become concave on the lower side. The cells of this palisade parenchyma respond much more actively to the stimulus of the fungus than do the subjacent cells of the spongy parenchyma, thus producing the curling. Link et. al. (1937) have reported that a substance having the properties of an auxin and which was presumably indoleacetic acid was extractable from *Taphrina deformans* culture filtrates. The role, if any, that this substance plays in the development of the disease picture as described above is as yet not clear. It is nevertheless true that auxins are capable of stimulating cell enlargement of the type described above in parenchymatous cells of many different plant species.

3. *Frenching and Shoestringing*. Frenching, a well-known deformity of tobacco, occurs in most tobacco-growing regions of the world. This condition is, in an advanced stage, characterized by a cessation of terminal bud and stem growth, and by a reticular type of chlorosis in the slowly expanding new leaves. This chlorosis may disappear as the young leaves develop and become strap-shaped (sword and string) as a result of the failure of the leaf lamina to expand. As apical dominance in such diseased plants is lost, the auxiliary buds develop and an unusually large number of leaves—which may be as high as 300—appears on a plant. These leaves assume an upright growth habit and are commonly sword- or string-shaped. Such plants have the appearance of a rosette or, in extreme instances, of a witches' broom. Root growth is also somewhat in inhibited, although the effect on the leaves appears to be the most characteristic feature of the disease.

The extent to which frenching modifies the morphology of a leaf can be seen from some figures that are given by Wolf (1935). Normal tobacco leaves average 55 cm in length and 30 cm in width. The corresponding measurements for sword leaves and string leaves are 40 cm × 10 cm, and 27 cm × 1.8 cm, respectively. Histological studies show, moreover, that the diseased leaves are two to three times as thick as normal leaves. This increase in thickness is due to an enlargement of cells of all the leaf tissue except cells of the vascular system. Schweizer (1933) has indicated that the xylem is markedly reduced in development in frenched leaves. In mature string leaves, the parenchyma remains quite juvenile with little or no evidence of dorsiventrality.

Although the cause of frenching is not clearly established, it is now believed to be a toxicity disease rather than the result of a nutritional deficiency. McMurtrey (1932) described briefly the similarity of frenching symptoms to those of thallium toxicity in tobacco. Spencer (1935) found that one part of soil obtained from a field in which frenching occurred mixed with 2,000 parts of sand produced typical frenching in the test plants. Thus, a toxic factor, effective in very low concentrations, was present in frenching soils. Spencer (1935) tested 33 different elements on tobacco and found, as had McMurtrey, that only thallium at a concentration of 5 p.p.m. or less produced chlorosis, strapshaped leaves and other symptoms characteristic of frenching. Later, Spencer and Lavin (1939) indicated that frenching and thallium toxicity are probably two distinct physiological conditions.

Steinberg (1950) has suggested that frenching of tobacco is caused by the action of diffusates from the presumably nonpathogenic soil bacterium *Bacillus cereus*. The effectiveness of *B. cereus* diffusates in eliciting typical frenching symptoms in tobacco varied with the kind and quantity of peptone used and the concentration of inorganic nitrogen in the test medium.

Bacillus cereus is a widely distributed soil microorganism. The progressive development and type of symptom produced by it in tobacco (in aseptic culture) largely parallel those that appear in plants subject to this abnormality under field conditions. Moreover, rhizosphere and rhizoplane counts of this organism increased by 65% and 200%, respectively, when frenching occurred in the field. The nature of the toxic substance is not yet known.

Steinberg (1952) has shown that slight excesses of certain amino acids caused production of symptoms resembling frenching in tobacco seedlings. Frenching symptoms were, however, limited to the natural isomers of alloisoleucine and isoleucine in tobacco and to these compounds and leucine in *Nicotiana rustica*. Alloisoleucine was most effective in both species and as little as 2 to 8 p.p.m. resulted in chlorosis and strapping of leaves in tobacco. Leucine, which was ineffective when applied to tobacco, was more effective than isoleucine in *N. rustica*. Analytical studies indicated, moreover, that frenching was found to be accompanied by a marked increase in isoleucine and other free amino acids in the leaf lamina of field-grown plants. The absence of free amino acids in frenching soil and the marked increase of free isoleucine in frenched leaves indicate that *Bacillus cereus* toxin and isoleucine are not identical. The conclusion was therefore drawn that the accumulation of excessive quantities of free amino acids in the strapped leaves was a probable chemical factor involved in the production of the morphological symptoms in the plant. The stages suggested by Steinberg (1952) leading to abnormal growth in frenching of field tobacco were, therefore, as follows: bacterial soil toxin → receptor → excessive accumulation of isoleucine and other free amino acids in the leaves → frenching.

A condition resembling frenching may also be caused by certain strains of the tobacco mosaic and cucumber mosaic viruses or a mixture of these viruses, as well as by a gene mutation. Kunkel (1954) has reported an extreme instance of shoestringing caused in tomato plants by a mixture of tobacco mosaic virus and the virus of carrot yellows of Texas. Leaf-blade development in this instance was completely suppressed. Just how these viruses suppress or prevent leaf-blade growth is not understood.

The so-called "wiry" tomato plants described by Lesley and Lesley (1928) are not unlike those which result from the virus diseases of the shoestring type reported above. Wiry plants appear, however, to result from genic mutation. This condition is recessive to the normal and the plants are completely sterile. The leaves of these plants are variable in shape but have a strong tendency toward reduction of the leaf lamina and in extreme instances the leaves consist merely of a tapering midrib.

C. Growth Changes in Stems and Branches

1. *Organs Arising in Unusual Places. a. Adventitious Roots.* The development of adventitious roots on stem of plants has been reported to be characteristic host response associated with certain bacterial, fungal, and viral diseases.

A number of investigators have described this phenomenon following infection of tomato and certain other hosts with *Pseudomonas solanacearum*. Smith (1920) reported it to be associated with the Grand Rapids disease of tomato which is caused by *Corynebacterium michiganense*. Adventitious root formation was found to occur on tomato and *Kalanchoe* following inoculation of those host species with the crown gall bacterium. In has also been observed by Wellman (1941) and Dimond and Waggoner (1953) in tomato plants diseased with *Fusarium* wilt. Certain viruses, such as the cranberry false blossom virus in tomato as well as that implicated etiologically in the sereh disease of sugar cane, characteristically stimulate root formation in the stems of their respective hosts.

Grieve (1940) studied the development of adventitious root formation in tomato plants infected with *Pseudomonas solanacearum*. This investigator found that the new roots arise commonly over a span of several internodes along the path of the primary bundle except in those instances in which the disease progresses with great rapidity, in which case these adventitious structures do not develop. Histological examination of transverse and longitudinal sections of infected plant stems demonstrated that adventitious roots usually develop in regions where large primary bundles are affected. The initiation of root primordia often precedes the advancing bacteria, indicating action at a disease. Grieve applied indoleacetic acid to the stems of tomato plants and found that the roots initiated were similar to those resulting from infection. This finding suggested that the bacteria are inducing root formation either directly as the result of the production of a growth-promoting substance, or indirectly through their interference with the metabolism of the plant. In attempting to distinguish between these two possibilities, Grieve found that *Pseudomonas solanacearum* produces indoleacetic acid from tryptophan in culture. However, virulent and avirulent strains produce approximately equal amounts of this substance. The possibility that the bacteria produce an auxin in the xylem by acting on naturally occurring or artificially introduced tryptophas was examined but no evidence for such production was obtained. No difference in auxin concentration could be detected, moreover, between healthy and diseased plants by bioassay methods. Furthermore, adventitious roots could be produced by. cutting or blocking the bundles. These findings led Grieve to question whether the bacteria induce the formation of adventitious roots directly through the elaboration of indoleacetic acid. He suggested that the production of these structures more likely results from disturbances of normal auxin transport in the plant as a result of mechanical blocking of the vessels by the bacteria.

Price and Gainor (1954) reported a striking correlation between adventitious root development and the number of leaves present on a *Kalanchoe* plant inoculated with crown gall. These workers suggested that a substance responsible for stimulation of adventitious roots is organized in leaves and is transported downward to the site of the tumor. That the biologically active substance is not indoleacetic acid was indicated by the finding that 2% of this substance in lanolin—applied to stumps of petioles above the points of inoculation—did not stimulate the development of adventitious roots in *Kalanchoe daigremontiana*.

De Ropp (1947) has shown, moreover, that a powerful root-stimulating substance is elaborated by sterile crown gall tumor cells. In crown gall it appears likely, therefore, that the stimulus for adventitious root formation originates in the cells of the host rather than directly from the causal bacteria.

The course of *Fusarium* wilt of tomato is not unlike that of the bacterial wilt of this host as described above. Wellman (1941) reported adventitious roots associated with this fungus disease. Since epinasty of the leaf petioles often appears on plants showing adventitious root formation, Dimond and Waggoner (1953) have suggested that in the case of *Fusarium* wilt, adventitious root formation, like epinasty, may result from the production of ethylene either by the causal fungus or by the cells of the host as a result of the interaction of the host and the pathogen.

b. Adventitious Shoots. The witches'-broom virus in potato causes the infected plant to produce numerous buds at the nodes in the above-ground stems of potato plants. Long slender stolons that resemble aerial roots but which are covered with trichomes develop from these adventitious buds. Aster yellows virus and carrot yellows virus from Texas, on the other hand, stimulate the production of small aerial tubers in the axils of the leaves of potato plants.

An extreme example of adventitious shoot formation was reported about 100 years ago by von Martius. This worker described an unusual plant that possessed a mania for forming innumerable leaves and shoots. This species has been appropriately named *Begonia phyllomaniaca.* Plantlets develop spontaneously in incredible numbers from the superficial cell layers of the leaf lamina, petioles, and stems. Erwin Smith (1920) studied phyllomania and believed it to be conditional on shock such as is encountered during the repotting of plants. Smith indicated, moreover, that the cells of leaves and internodes are susceptible to such shock only during a relatively brief period of meristematic growth. The adventitious shoots do not arise from preformed buds but develop from totipotent cells at the base of the trichomes and especially from botryose glands which are found in great abundance in young stems and leaves of this species. These embryo plants develop a vascular system of their own but the vast majority never succeed in connecting this with the vascular system of the plant. They must, therefore, be considered not as branches but rather as independent organisms.

2. *Witches'-Brooms.* Witches'-brooms, or "hexenbesens", are closely grouped, much branched structures commonly found to occur on a number of different species of trees and shrubs. The stimulus necessary for their formation is supplied in different plant species by pathogens of the most diverse types. Fungi of the genus *Taphrina* and various rust fungi are effective in inducing witches'-broom, while distinct virus species have been shown to cause the formation of such an abnormality in alfalfa, potato, peach, and the black locust. Bos (1957) has recently presented a detailed account of the virus-induced witches'-broom. An eriophyd mite belonging to the genus *Aceria*, followed by a powdery

mildew fungus, is said to be involved in the development of branch knot so commonly found on the hackberry.

Typical witches'-brooms caused by the fungi appear to live a more or less independent life and act as parasites on the plans from which they are derived. In accordance with their independent existence, witches'-brooms tend to break away from the correlations of the parent plant. Instead of branching out horizontally, the brooms stand as more or less erect clusters of branches. A normal dorsiventral symmetry is thus changed to a radial symmetry. Witches'-brooms as a rule do not produce flowers, indicating further a breakaway from the correlations of the parent plant. Heinricher (1915) has shown, moreover, that a twig of a witches'-broom caused by *Taphrina cerasi* grafted to a healthy sweet cherry tree develops again into a typical witches'-broom. The independence of these structures is further shown in an impaired periodicity. The vegetative buds found on the brooms of the sweet cherry, for example, open several weeks earlier than do those present on healthy branches. Gäumann (1946) has pointed out that this premature unfolding of the buds is probably associated with the fact that the shoots and buds comprising the broom never achieve a true winter dormancy since the pathogen never becomes entirely quiescent. The incomplete winter dormancy may result in a winter killing of the first year twings present on a broom.

The witches'-broom of the silver fir caused by *Melampsorella caryophyllacearum* is typical of this peculiar type of growth. The primary infection is said to occur in the young bark of branches surrounding buds. In the spring when the buds develop, the fungus mycelium grows into the epidermis of the developing shoot and penetrates the cortex so that by fall a slight swelling of the shoot axis is found. During the following years an overgrowth of considerable size may be formed and buds embedded in this growth develop to produce the characteristic deformation. The twigs of this particular type of broom are found to develop in whorls. They are short, thick, soft, and pliable. This results from the fact that cortical parenchyma is spongy and the wood is not well developed. The buds on the broom open earlier in the spring than do those present on a healthy twig, while the leaves found on the diseased specimens remain short, are yellowish in color, and fall off when a year old. The leaves present on a normal twig are, on the other hand, long, straight, dark green on the upper side, and commonly remain in position 5 or more years. The longevity of the diseased twigs themselves is limited and they die within a few years, giving rise to the dry, bristling brooms characteristically found on the silver fir.

The witches'-broom-like effects that occur in certain plant species following virus infection appear to result from the excessive stimulation and development of secondary shoots. This condition is not accompanied by swellings or overgrowths characteristic of many of the fungus-induced brooms.

Brian (1957) has suggested that the diseases characterized by excessive branching may well be due to auxin deficiency rather than to auxin excess. Lacey (1948) has, in fact, shown that cultures of *Corynebacterium fascians*, the causal agent in leafy gall development on certain plants, are capable of rapidly destroying indoleacetic acid and other auxins present in plant tissues.

D. Growth Changes in Floral Organs

1. *Alterations of Floral Parts*. Under the stimulus of pathogens, the sepals, petals, stamens, or pistils of a flower may be transformed into structures that are very different in appearance from those found normally. Kerner von Marilaun (1891) has reported that double flowers are produced in *Valerianella carinata*, the common corn salad, as a result of infestation by a mite. This doubling results from a retrograde alteration of the stamens into a whorl of petals. These petals, under the stimulus of the mite, enlarge to more than fifty times their original size and finally appear as fleshy lobes which fuse with one another into a disc. The greatly enlarged lobes bend backward and are concave on the lower side. It is in the cavities thus formed that the gall mites live.

In the capitula of certain of the milfoils, *Achillea millefolium* and *A. nana*, the peripheral ray florets and the central tubular ones become leaf-like in appearance and assume remarkable forms as a result of mite infestation. A capitulum is often subdivided into several stalked sub-capitula, while the flowers are altered into green funnel-shaped structures with jagged mouths and into small flat-lobed green foliage leaves, while short green scale-like leaflets, which represent modified stamens, develop from the midribs of these leaves. Sometimes, however, the changes in growth pattern following mite injury are not so extreme as the two instances cited above might suggest. In *Veronica saxatilis*, for example, the development of numerous hairs on the rachis of the raceme, the pedicels, and the bracts appears to be the only characteristic host response. Hairs are not present in the corresponding normal structures of this plant species.

Doubling of the flowers of the alpine rose, *Rhododendron ferrugineum*, has been reported by Kerner von Marilaun (1891) to result from gnat infestation. In this instance the stamens and carpels are transformed into red petals. Since flowers of this species normally have 10 stamens and 5 carpels, there should be only 15 red petals in the center of each but there are often two or three times that many present. It thus appears that not only metamorphosis but multiplication takes place.

According to Tschirch (1890), the aphid *Astegopteryx stryaeophila* induces remarkable changes in young flowers of *Styrax benzoin*. The calyx, corolla, and androecium are transformed into large abnormal leaves that form bag-like pockets. The pistils, however, appear to be unaffected in this instance.

Metamorphosis of floral parts of higher plants are also brought about by certain species of fungi. The transformation of stamens into petals commonly follows infection of *Viola sylvestris* by *Puccinia violae*. Similarly, *Peronospora violacea* has been reported to encourage stamen primordia in the flowers of *Knautia arvensis* to develop into petal-like structures. An extreme instance of phyllody is found when flowers of the Japanese plum are infected with a rust of the species *Caeoma makinoi*. In this instance all of the floral parts are transformed into foliage leaves. In the alder the bracts of the pistillate flowers are changed by *Taphrina alni-incanae* (*Amentorum*) into greatly elongated purple-red spatulate lobes which are twisted and bent.

Von Tubeuf (1895) has reported that when *Albugo candida* infects the inflorescences of the radish, the ovary, calyx, and corolla enlarge and the androecium assumes a leaf-like appearance. Stamens may become green and leaf-like in the downy mildew disease of *Pennisetum glaucum* and other grasses. Thus, this disease, which is caused by *Sclerospora graminicola*, is popularly known as green ear. In the case of head smut of maize, caused by *Sorosporium reilianum*, the whole staminate head or the ear may become a leafy structure.

In the United States, there are at least three different viruses, the aster yellows virus, the cranberry false blossom virus, and tomato big bud virus, that produce gigantism in the floral organs and more particularly in the sepals and calyxes of tomato and certain other solanaceous plants. The tomato big bud virus appears to affect the sepals rather specifically. These structures enlarge greatly under the influence of this virus and fuse to form huge bladder-like growths that conceal the inner parts of the flower. The stamens and pistils do not appear to be greatly affected. The aster yellows virus and the cranberry false blossom virus cause symptoms in the flower trusses of tomato that are almost identical with those caused by the big bud virus. Although these three viruses cause the sepals of certain of the solanaceous plants to enlarge greatly, they do not have this effect on sepals of species of certain other families. They do not, for example, cause gigantism in the sepals of the flower trusses of the periwinkle, *Vinca rosea*, but they bring about a virescence in the petals, stamens, and styles. There is, in this instance, a retrograde development of floral parts into foliage leaves. It is, therefore, clear that the type of symptom produced depends as much on the species of plant in which the virus multiplies as on the nature of the virus.

The aster yellows, cranberry false blossom, and carrot yellows virus of Texas stimulate the formation of adventitious buds in the stigmas of flowers of a number of different hosts. Such buds may give rise to either flowers or to leafy stems which in turn may bear flowers.

Bos (1957) has recently attempted to interpret the development of the flower and its component parts on the basis of the phenomenon of antholysis accompanying virus infections in *Crotalaria* and certain other hosts. This investigator has suggested, on the basis of the homology of the reproductive and vegetative parts of a plant and their morphogenetic development, that the sexual and vegetative activities in the plant are mutually antagonistic. A growing point can develop only into an inflorescence or into a vegetative shoot. During flower initiation the vegetative characters are suppressed and sexual characters prevail. The sexual characters remain suppressed, on the other hand, during vegetative growth. From the manner of appearance of antholysis in virus-infected *Crotalaria* plants, Bos has concluded that flower induction is stopped suddenly and that subsequent development of the floral parts proceeds exclusively in a vegetative manner. The ultimate result obtained depends upon the stage of development of the primordia in the bud at the time of suppression of flower induction by the virus. Thus, flowers initiated in succession, produce a series of flowers showing increasing antholysis. These—in reverse order—clearly show a macroscopically recognizable picture of the morphogenesis of the flower and of its component parts. Therefore, according to Bos, antholysis supports the theory that the flower must be regarded as a modified leafy branch.

2. *Overcoming of the Normally Arrested Development of Floral Parts.* The anther smut, *Ustilago violacea*, produces its spores only on the anthers of certain hosts. When this fungus infects the female flowers of

Melandrium album or *M. dioicum*, the stamens, which are normally arrested in their development and are present only in a rudimentary form in the pistillate flowers, grow to full size but when mature are filled with smut spores instead of pollen.

A similar process is said to occur when pistillate flowers of *K. arvensis* and *K. sylvatica* are infected with *Ustilago scabiosae*. The nature of the morphogenetic stimulus which overcomes the arrested development of male sex organs in female flowers is still unknown.

E. Modification in Fruiting Structures

Modification of the fruiting bodies of higher plants occurs quite commonly as a result of infection. Such descriptive names as little peach, bladder plum, phony peach, etc., have been applied to conditions of this type. Hypertrophied fruits are commonly produced in certain species of the genus *Prunus* as a result of infection by *Taphrina pruni*. In this instance, the tissue of the diseased ovary is stimulated to growth, but not in the same way as in the normal fruit. The resulting body is flattened on two sides, is brittle, yellow in color, and much longer than the normal fruit. The seed within is abortive and a hollow space is left in its place. These hypertrophied growths, which are commonly called "bladder plums", fall from the trees at the end of May and are said to be eaten in certain areas. A somewhat similar condition is found when *Taphrina durea* infects the pistillate flowers of the poplar. This fungus, like the one described above, stimulates the growth of the ovaries with the resulting development of golden yellow capsules that are more than twice the normal size. The smut fungus *Ustilago zeae* also stimulates growth of the tissue of the pistillate flowers of maize. As a result, the grains are replaced by irregular cushion-like structures with a diameter of up to 7 cm. The resulting growths contain more auxin than do normal tissues. *Ustilago zeae*, moreover, produces an auxin in culture which was identified as indoleacetic acid by Wolf (1952). The fruit size may be greatly dwarfed as in the case of two virus diseases, little peach and phony peach, or the fruits may assume an asymmetrical shape as is found characteristically in the xyloporosis disease of oranges—which is also of viral origin.

The seed capsules of *Datura stramonium* normally bear numerous conspicuous spines. When plants of this species are infected with the severe etch virus, spine formation may be completely inhibited. The aster yellows virus, on the other hand, causes a bursting of the abnormal fruits produced by *Cajophora lateritia* plants infected with this agent. The seed-like structures present in these fruits are green in color and have been transformed, under the influence of the virus, into short stems which bear leaves.

The entire inflorescence may, in certain disease conditions, be altered to quite a different type. The rust fungus, *Aecidium esculentum*, is said to cause the normal inflorescence in *Acacia* to change from a head to a spike. Similarly, the stinking smut fungus, *Tilletia tritici*, may change the growth pattern of the head of the club types of wheat to the elongated or "vulgare" type.

Individual floral parts or even entire organs may be replaced by new structures in certain diseases. *Claviceps purpurea* invades and destroys certain of the ovaries of rye and other grasses. In place of the seeds, elongated dark purple sclerotia are produced. These are composed of dense aggregates of fungus mycelium.

In sheep sorrel infected with *Ustilago oxalidis* the seeds are replaced by the spores of pathogen. In this instance, the spores are forcibly expelled from the seed capsule as if true seeds were present.

F. Fasciation

A condition known as fasciation can be classified between that group of growth abnormalities described above (involving harmonious changes in growth pattern) and the amorphous changes to be considered later in this discussion. Fasciation is a morphological term that has been used to describe a series of abnormal growth phenomena resulting from many different causes, any of which result at the morphological level in a flattening of the main axis of the plants. Although this ribbon-like expansion of the stem is often the most striking feature of this condition, all parts of the plant may be affected. Fasciation often results in alterations in the arrangement of foliar and floral structures. White (1948) points out that, when fasciation occurs, the early seedling growing stages are normal. As the plant develops, however, the growing point becomes broader, and the unregulated, distorted tissue growth results in significant increases in weight and volume of plant tissue. The apical growing region becomes linear and comb-like in some instances or develops numerous growing points, producing a witches'-broom effect. In still other instances, the growing points may be coiled and resemble a ram's horn or they may be highly distorted into a grotesque tangle of coils. Fasciations are widespread, both geographically and taxonomically. Examples have been recorded in 102 families of vascular plants. Fasciations found in certain plant species such as the common cockscomb,

Celosia cristata, as well as the cristates in the cacti are highly prized by gardeners.

Fasciations have been classified on morphological grounds—based on such physical features as form, color, and anatomical structure. Linear, bifurcated, multi-radiate, and ring or annular types have been listed. This condition has also been classified from a causal standpoint. In this type of classification the conditions and agencies necessary to produce the character as well as the modifying factors that affect its development are considered. White (1948) has placed these essentially into five categories (1) fasciations which breed true and which in crosses with the normal obey genetic laws so that their genic basis is known; (2) noninherited forms in which the fasciation is due to environmental causes and the character is not reproduced in the selfed seed; (3) fasciations that occur spontaneously and which have been propagated vegetatively but in which neither the initial cause nor the question as to whether the condition is transmitted through the seed is known; (4) fasciations which are induced artificially by known procedures; (5) fasciations which have been imperfectly investigated.

Fasciations resulting from modifications of the normal gene complement, and which breed true, have been known for centuries in such plants as *Celosia* as well as in the mummy or crown type of garden pea. Gregor Mendel (1866) found fasciation in the mummy pea to be recessive in the F_1 when crossed with normal varieties. Such F_1's showed a ratio of 3 normals to 1 fasciate in the F_2. A gene-fasciated race in *Nicotiana tabacum* when crossed with nonfasciated strains, on the other hand, gives 1 : 2 : 1 F_2 ratios. In other instances, very complicated results were obtained, showing the effects of modifying factors and giving rise to nonfasciated segregates which carry the fasciated gene. Fasciation is not correlated with changes in chromosome numbers but it may produce meiotic irregularities.

Aside from gene mutation, fasciation has been reported to result from environmental effects such as frost, pressure, alteration of food and water relationships, pruning, mutilation, as well as by more specific etiological agencies such as a bacterium, *Corynebacterium fascians*, and X-irradiation.

There seems little doubt but that nutritional changes due to correlative disturbances in growth-substance relationships play a role in fasciation. Bloch (1938) has reported that this condition can in certain instances be produced experimentally by treating plants with auxincontaining pastes.

On the other hand, high doses of X-rays have been result in fasciation. The effects of X-irradiation on auxin metabolism in the growing plant are obviously complex. These nongenetic types of fasciation are not transmitted through the seed but may, in some instances, be propagated asexually. Both inherited and noninherited types of fasciations which are phenotypically indistinguishable have been described for at least 3 different plant species.

Orland E. White believes that the basic cause of fasciation is a disturbed metabolism involving excessive nutriment which mobilizes energy that must be used. This energy, once accumulated, goes, according to White (1948), into growth and becomes "wildly" expended in abnormal and unpredictable tissue production. Heslop-Harrison (1952) points out, however, that these conclusions are descriptive rather than truly explanatory since excessive nutrition in itself cannot account for the abnormal element in fasciation. He believes, rather, that a proximate cause of these abnormalities is probably to be found in maldistribution of the auxins in the plant. Jones (1935) and Orland White (1948) have suggested that fasciation bears an analogy to animal cancer. In the opinion of the writer, however, this condition does not represent a true tumor at all but might better be placed in that category of teratological abnormalities known as monstrosities.

II. AMORPHOUS CHANGES IN GROWTH PATTERN

Amorphous changes involving either the temporary or permanent loss of typical organization are found to occur very commonly in plants. These overgrowths range from somewhat exaggerated but self-limiting wound-healing responses at the one extreme, to rapidly growing nonself-limiting tumors which have no characteristic size or structure, at the other.

A. Self-Limiting Overgrowths

1. *Intumescences.* Intumescences, which are among the simplest and most innocuous of this type of growth, are found to occur most commonly on leaves but are also found on stems and fruits of plants. These blister-like pustules usually result from the abnormal elongation of groups of cells with or without increased cell division. In a typical leaf intumescence found in many plant species, the palisade parenchyma cells elongate considerably in localized areas to give rise to the pustules. Intumescences have been found to be produced experimentally in a

number of ways. No pathogens, however, appear to be concerned in their development. Sorauer (1886) believed them to result from an excess of moisture in the air and soil. Harvey (1918) produced intumescences artificially on the undersurface of cabbage leaves by subjecting such leaves to slight freezing. Smith (1920), on the other hand, induced blister-like pustules in cauliflower leaves with the use of such irritating chemicals as formic and acetic acids. Wolf (1918) has shown that intumescences may be produced in cabbage leaves by means of a sandblast and this worker has indicated that such overgrowths found naturally in the field result from sand driven against the leaves by wind. More recently, La Rue (1935) was able to reproduce perfectly the intumescences which arise under conditions of high humidity on the leaves of poplar with the use of low concentrations of indoleacetic acid. Much larger self-limiting over-growths were produced artificially by Brown and Gardner (1936) on bean plants with the use of higher concentrations of that substance. It is not unlikely that certain of the burls as well as the overgrowths that sometimes arise at graft unions result from a hormonal imbalance of this type.

2. *Galls*. Localized overgrowths in which the host cells are stimulated to excessive growth by pathogens are known as galls. In these instances, the continued growth of the host cells is dependent upon continued stimulation by the pathogen. There are unusually large numbers of distinct overgrowths of this type to be found in plants of which only a few representative examples have been selected for discussion.

a. Insect Galls. Among the most interesting types of self-limiting growth abnormalities found in plants are those that result from the activity of certain of the gall-forming insects. The whole subject of insect galls as reviewed in detail by Küster (1911), Ross and Hedicke (1927), and more recently by Felt (1940) suggests that many of these highly specialized overgrowths represent beautiful examples of dependent differentiation.

Insect galls may result either from a mechanical or chemical stimulus. Stem swellings on roses caused by the closely placed spiral galleries of the rose stem girdler, *Agrilus viridis*, are believed to be an example of the former type. Chemical stimulation, however, is probably far more important than mechanical irritation in the production of many insect galls, although the chemical stimulus may in some instances be supplemented by directive feeding of the insect. An examination of the literature leaves the unmistakable impression that highly specific morphogenetic stimuli, of chemical nature, and elaborated by insects, are capable of initiating, stimulating, and directing most precisely the development and differentiation of plant cells. There is no question about the fact that the morphological form that a gall assumes depends upon the nature of the pathogenic insect. This is evidences by the fact that the same host species or even the same organ of the same plant attacked by different but closely related insect species produces morphologically very different galls. Figure 4.2 illustrates this point and shows drawings of 4 galls of quite different morphology induced on leaves of the California white oak by 4 closely related species of cynipids. The morphology

Fig. 4.2. Four morphologically distinct galls produced on leaves of the California white oak by four closely related species of insects.

of the gall is so specific that it is considered by some to be a more reliable criterion for distinguishing between closely related species of insects than are the morphological characters of the adult insects. The morphology of the gall does not, moreover, appear to be a function of the part of the plant from which it arises. Current galls of the oak, which are found on both flower stalks and leaves, have similar morphological structure at both points of origin. The same insect species may, moreover, produce the same gall type on different species of plants. The sawfly,

Micronematus gallicola, produces bright red galls of similar morphology on four different species of willow. Findings such as those reported above suggest in the strongest possible manner that a large number of highly specific chemical substances are elaborated by insects and that these substances have specific morphogenetic effects on the cells and tissues of a plant. However, only moderate success has thus far been achieved in establishing experimentally the truth of this view. That a morphogenetic stimulus is elaborated by certain insects is suggested by the fact that the Norway spruce aphid, *Adelges abietis*, attaches itself firmly by its beak to the scale of a fir bud and can directly influence only a few cells of the young shoot. Nevertheless, thousands of cells on this shoot are affected and soon begin to assume an altered form and are stimulated to growth, indicating that a diffusible stimulus is active. Plumb (1953) has, in fact, induced the formation of this gall by injecting a glycerol extract of the salivary glands of the insect into the bases of spruce buds. Beijerinck, as early as 1888, killed the eggs of a sawfly with a hot needle immediately after they had been deposited in the host and found that the galls developed normally. Beijerinck called the biologically active materials "growth enzymes", and in 1897 applied this concept to the development of form in general. Parr (1940) showed that the gall-forming coccid, *Asterolecanium variolosum*, stimulates the host tissue to both hyperplastic and hypertrophic activity. He demonstrated further that the effect of the salivary secretion of this homopteron continues to stimulate plant cells even after the insect itself is gone. It was possible to reproduce galls similar to those produced by the insect by injecting glycerol extracts of the salivary glands. When the salivary extracts were heated to 60°C, they did not produce galls, indicating that a heat-labile, enzyme-like substance was involved in gall formation.

Rössig (1904) investigated various organs of the gall wasps and believed that he could trace the site of formation of the gall-forming substances to the Malpighian vessels. Similar findings were reported by Triggerson (1914) for the cynipid which produces the oak-hedgehog gall. Lewis and Walton (1947) reported some very interesting histological and cytological findings dealing with a biologically active substance believed to be involved in the growth and differentiation of cell development of the cone gall of witchhazel. This gall is produced by the aphid species *Hormaphis hamamelis* Fitch. In this instance the stem mother inserts her stylets into and between the cells of an immature leaf and injects very minute droplets of a substance secreted by glands opening into the stylar canal. This stinging process is not to be confused with feeding and the sting substance is quite different in its nature from substances found in the salivary secretions. When about 150 stings have been made in a small circular area by the stem mother, the cells in this region at first become etiolated. Almost immediately thereafter, cells receiving injected sting material dedifferentiate, undergo rapid mitotic divisions, and then redifferentiate into gall cells rather than typical leaf cells.

The injected sting material consists of a ground substance in which minute crystalloids are embedded. These crystalloids show a reddish purple cast in such stains as gentian violet, Congo red, and acid fuchsin. They are capable of passing readily from cell to cell as well as of entering cells when they are deposited by the insect in the intracellular spaces. Whether the injection is inter- or intracellular, the end result is the entry of the crystalloids into the nucleolus. There they may fuse to form a large crystalloid which again breaks up into smaller ones as mitosis begins. During nuclear division, the crystalloids are distributed to the daughter nuclei where they are again found in the nucleolus. Since these bodies are incapable of self-propagation and since they are apparently used up during growth of the cells, fresh sting material must be injected repeatedly by the stem mother during the entire growth of the gall. It would indeed be interesting to know the chemical nature of the "crystalloids" since they possess some very unusual regulatory properties.

J.P. Martin (1942) induced galls on three-month-old sugar cane plants by injecting extracts obtained from macerated leaf hoppers of the species *Draculacephala mollipes*. These findings again appear to implicate hormone-like substances in the initiation and development of insect galls. Anders (1958) reported that he was able to reproduce the swellings associated with the *Phylloxera* disease by applying to grape plants secretions obtained from the aphid responsible for this condition. A similar type of swelling was obtained when certain amino acids, used in the same proportions found in the insect secretions, were applied to grape roots.

The larva of the moth, *Gnorimoschema gallaesolidaginis*, induces an eliptical monothalmous gall on the stem of the *Solidago* host. The larva in this instance burrows into the terminal bud of the plant and then down the stem to a point 2 cm below the growing point, where it eats away the central tissues and induces the formation of a simple spindle-shaped gall. The gall stimulus, according to Beck (1954), appears to be associated with a silky substance secreted by the feeding larva. This silky material induces anatomical changes in normal stems similar to those found in the galls. The stimulus is, however, short-lived and uniform and continued deposition of the silky substance over the surface of the larval chamber appears necessary for the formation of typical galls.

Boysen-Jensen (1952) studies, with the use of rather ingenious methods, the development of a midge (*Mikiola fagi*) gall on beech leaves. Evidently the formation of the gall in this instance is caused by growth substances given off by the larva. These as yet uncharacterized substances produce cell enlargement and cell division but not organized growth. It was, therefore, suggested that cell enlargement and cell division are regulated by the larva which moves rapidly about the gall chamber and secretes the growth-promoting substances in definite places, thereby making the gall assume its special form. According to this interpretation, the growth-promoting substance does not have special organizing properties but the shape of the gall is dependent upon the distribution of a rather nonspecific type of growth substance by the larva. The growth substances are, in other words, tools which are used by the gall larva to model a gall from the cells of a beech leaf. While such a mechanism as that proposed by Boysen-Jensen may satisfactorily explain the formation of certain simple galls, it is difficult to see how it would explain the development of the more intricate structures the cells of which are highly differentiated.

The distinguishing feature of insect galls in general, and the more highly developed cynipid galls in particular, is the determinate growth of these structures. Bloch (1954) has suggested that an insect gall is almost comparable—in its "determinate" growth—to a leaf or a fruit. These galls are constant in form and size and possess their own polarity and symmetry. Although the cells of these galls dedifferentiate as a result of the initial stimulus, they again redifferentiate into an orderly rearrangement of cells layers which possess a degree of differentiation that is never below that of the host. A fascinating field lies open here for exploration at the morphogenetic and biochemical levels.

b. Root Nodules. The root nodules found to arise on many species of leguminous plants as a result of infection by bacteria of the genus *Rhizobium* are, like the insect galls, highly organized and specialized structures. These nodules, unlike most of the overgrowths described here, are by and large highly beneficial to the host because of the role that they play in nitrogen fixation.

The typical root nodule is composed of four histologically well defined regions. The outermost tissue, or nodular cortex, consists of several layers of parenchymatous cells that originate from the nodular meristem found immediately below the cortex at the distal end of the nodule. The meristem is conspicuous and is composed of small, compact, rapidly dividing cells. The cortex and meristem are normally free of bacteria. The provascular tissue of the nodule arises as a result of radial divisions and differentiation of cells—at the periphery of the inner infected cells and the nodule cortex—at the time when the nodule is still meristematic. These differentiate later into a typical vascular bundle which consists of xylem, phloem fibers, sieve tubes, and companion cells enclosed in parenchymatous tissue and surrounded by an endodermis. The vascular system of the nodule is connected with that of the host and is functional. The central region of the nodule is composed of two types of parenchymatous cells, infected and noninfected, and is commonly known as the bacteriod zone.

Essentially, two types of nodules have been recognized on the basis of their origin. The more common or exogenous type arises, except for the vascular linkages, from the cortical parenchyma of the root. The second or endogenous type is found far frequently and is composed of cells that arise from proliferation of the pericycle.

The information presented above concerning structure clearly indicates that the nodule is not a shapeless mass of cells but instead represents a highly organized growth. This information, in turn, permits a consideration of the possible mechanisms involved in the growth and development of the nodule.

The infecting bacteria commonly enter the host through root hairs, although in certain aquatic plant species that do not possess root hairs they may and commonly do enter through epidermal cells. Prior to invasion of the host, the bacteria form a small colony close to the tip of the root hair. Under these conditions, the hair curls very characteristically and assumes the form of a "shepherd's crook." As early as 1900, Hiltner found that sterile filtered bacterial extracts, when applied to root hairs, produced a comparable type of deformation. Similar results were reported by Thornton (1936) and Thornton and Nicol (1936) who showed that such filtrates stimulated the production, growth in length, and characteristic deformation of the root hairs. Later, Thornton (1947) found that pure β-indoleacetic acid, when applied to root hairs, elicited a curling similar to that found prior to invasion by the bacteria. Probably, therefore, the characteristic curling of the root hair is the result of the production of an auxin by the bacteria although this points has not yet been fully established. Wilson (1940) points out, for example, that the characteristic curling of the hair is not easily explained on the basis of a growth hormone present in the sterile culture filtrate. He states that one would not expect to find differential growth rates under such experimental conditions since the concentration of the hormone should be the same on all sides of the roots.

After the bacteria enter a hair, they align themselves into hyphalike zoogloeal infection thread. The thread penetrates to the base of the hair and then into and directly through the subjacent cortical cells where it branches. As the zoogloeal stands migrate through the cortical cells, they become encased in a sheath composed of cellulose-like material, which on the basis of cytochemical studies is said to be composed of the same material as is the wall of the host cell. This has led to the suggestion that the sheath is deposited by the host cells rather than by the zoogloeal strand and that it serves as a defense mechanism against the invading bacteria. Of particular interest to the present discussion is the fact that proliferation of the root cells and subsequent development of the nodule do not occur until the bacteria are released from the infection thread. Release may be achieved essentially in the following way. The bacteria may be discharged from the tip of the thread as the thread migrates. They may be released as a result of the rupture of unsheathed globular protuberances that are commonly found to be present in infection threads or they may be liberated following stresses accompanying cell enlargement and cell division. Wipf and Cooper (1940) have suggested, on the basis of cytological studies, that invasion of the disomatic cells by the bacterial is essential for the release of the bacteria from the infection thread as well as for cellular growth terminating in nodule formation.

Once the free bacteria reach the cytoplasm, they tend to migrate to the peripheral regions of the cell and begin to multiply. The host cells containing the organisms, as well as the adjacent uninfected cells, undergo rapid division and subsequent enlargement. Infected cells may increase eightfold in size. Since neighboring uninfected cells are also stimulated to grow, it has been suggested by McCoy (1929) that the bacterial elaborate a diffusible substance which stimulates cell enlargement and cell division. Ultimately, many of the cells in the heart of the nodule become filled with bacteria and the rods found in nodules of certain, but not all, plant species become highly pleomorphic. It is in this so-called bacteroid zone of the nodule that symbiotic nitrogen fixation takes place. This zone also contains four highly interesting pigments, one of which, leghemoglobin, is closely related chemically to blood hemoglobin. The occurrence of the pigments in nodules and possible role in nitrogen fixation has been reviewed in detail by Allen and Allen (1954).

Attempts have been made throughout the years to explain in physiological terms the mechanism involved in the development of the nodule. As early as 1912 Molliard reported that he had produced tuberizations on roots of pea plants by treating the roots with cell-free filtrates upon which the rhizobia had grown. It was concluded from these experiments that the bacteria elaborated some biological active, growth-promoting substance which brought about changes in the root similar to those found during actual infection. Thimann (1939) examined this question further and postulated the following series of events after the bacteria had penetrated the root. During the course of their metabolism within the host cells the bacteria elaborate a small amount of auxin and certain other substances. The auxin causes an enlargement of the cells in which it is produced but, being diffusible, enters the pericycle and stimulates growth and division, thus giving rise to the early stages of a lateral root initial. In the presence of continued auxin production by the bacteria, the potential lateral root is prevented from elongating. Instead, the cells increase isodiametrically in size. Uninfected cells are stimulated to division by auxin diffusing from the infected cells. Fred et. al. (1932) have discussed the evidence for and against this concept in detail and conclude that the nodule is not a modified lateral root "for it has no central cylinder, root cap, or epidermis. Further, it does not digest its way out from the cortex of the main root but remains covered with a considerable layer of cortical parenchyma."

Nevertheless, growing nodules are active auxin-producing centres and that substance plays an important role in the growth of the nodule has shown, moreover, that auxin production in the nodule roughly parallels the growth of the nodule.

Chen (1938), Georgi and Beguin (1939), Link (1937), Rasnizina (1938), and Thimann (1939) have demonstrated that *in vitro* the root-nodule bacteria produce considerable quantities of auxin from tryptophan, yeast extract, and peptones present in an otherwise suitable culture medium. Hunt (1951) found, moreover, that free tryptophan was present in the nodules. An evaluation of these findings is complicated, however, by the observation made by Chen (1938) and Georgi and Beguin (1939). These workers found that both effective and ineffective strains of rhizobia produced auxin from tryptophan and in some instances the ineffective strains appeared to be more efficient growth-substance producers than were the effective strains.

While it has been possible to produce overgrowths on the roots of certain leguminous plants by applying growth substances of the auxin type, it nevertheless is true that histologically and anatomically these artificially induced growths do not bear the slightest resemblance to nodules produced under the stimulus of the bacteria. Allen and Allen (1954) point out that the nodule is not a shapeless mass of cells but is a well organized structure. Thus, while there appears to be no conflict on the question of the increased growth-substance content of the nodule, a suitable explanation regarding the role that such a substance plays in directing the organization of the nodule is not yet at hand.

c. Root Knot. Root knot, which is caused by species of nematodes belonging to the genus *Meloidogyne*, is a destructive disease in many cultivated and wild plants. The overgrowths produced on the roots may appear as small scattered tubercles or as extensive swellings which may reach diameters up to 2 in. and involve almost the entire root system. Severely infected roots have a rough, clubbed appearance not unlike that encountered in the club root disease of the crucifers. Once embedded in the tissue, the larva begins to feed and almost at once stimulates the growth of the host cells which leads to the development of the typical knot. The gall, in this instance, is composed of relatively few, greatly hypertrophied cells together with cells showing hyperplasia. Often, 6 to 20 nuclei are present in the enlarged cells although Nèmec (1910) has reported more than 500 in a single cell of *Vitis gongylodes*. The multinucleate giant cells appear early in gall development. They are rich in storage materials and serve to nourish the larvae in much the same manner as do the "nutritive cell" layers found in certain insect galls. It would appear likely that at least two distinct stimuli, one of which is concerned with cell enlargement and which involves only a few cells and the other with cell division, are involved in the development of this type of overgrowth.

Dropkin (1954) has recently shown a high positive correlation to exist between gall area and numbers of larvae in a gall. He therefore cocluded that the response of the root to the presence of the nematode is a local one and that the size of the gall ultimately produced is a function of the amount of stimulation provided by each larva present in the overgrowth. Here, then, is another example of a self-limiting growth, the development of which is dependent upon continued stimulation by the pathogen.

d. Clubroot. The typical overgrowth produced on the roots of cabbage by *Plasmodiophora brassicae* is spindle-shaped, thick in the middle, and tapering gradually toward the ends. Such a club is a morphological unit and is commonly the result of a single primary infection. Sometimes, however, the swellings resulting from two or more primary infections fuse to form a compound club which is more irregular in outline than those resulting from a single primary infection. Cunningham (1914) has recognized different types of hypertrophy in different species of susceptible hosts. In cabbage, a complete, clubbing of the main roots and of the lateral roots may occur, while in *Sisymbrium altissimum* the lateral roots are commonly free of distortion. In *Sisymbrium officinale* and *Erysimum cheiranthoides*, on the other hand, clubs develop on the lateral roots but not on the main root. Both lateral roots and the main root are affected in *Lepidium sativum*. In this instance, however, club-free rootlets are found above the diseased portions. In the radish, *Raphanus sativus*, clubs occur as tumors on the root.

As a result of the overgrowths, infected roots fail to absorb nutrients from the soil and are often unable to transport food and water taken up by healthy roots. This functional failure of the roots results in the starvation and, ultimately, severe stunting of affected plants.

The spread of the disease-producing agent from points of primary infection is accomplished in cabbage through a direct penetration of the host tissue by the infecting plasmodia, which then grows and divides repeatedly. According to Kunkel (1918), infection by penetration may be divided into four parts: (1) primary infection of the cortical tissue and penetration to the cambium, (2) infection of the cambium in all directions from the point of original penetration, (3) passage of the plasmodia out of the cambium into the cortex and in from the cambium toward the xylem region, and (4) infection of the medullary rays.

As the plasmodia pass through the tissues, some of them become established within cells, while others continue to penetrate into the deeper regions. Although no noticeable effect is observed in a cell if infection is temporary, stimulation leading to abnormal cellular growth and division results if the infection of a cell is permanent. The period of infection continues up to the time that the host plant stops growing or dies. The infected cells of a club are distributed in small groups throughout the diseased tissue and do not lie adjacent to each other. The stimulus resulting in abnormal cellular growth appears to travel in advance of the infection. This can be clearly seen when infection is established in the medullary rays. The ray cells—in the proximity of diseased cells—increase considerably in size. This increase in size as well as in number of cells tends to split the central cylinder of an infected root. The mass of parasitic protoplasm in a given volume of diseased tissue is, according to Kunkel (1918), remarkably constant in different clubs. This indicates that the amount of cell growth in this disease, as in the case of root knot, is dependent upon the mass of parasitic protoplasm in the diseased tissue. Clubroot can be considered to be a malignant disease in the sense that this malady may and frequently does kill plants that are affected by it. However, there is no evidence that the affected cells themselves have become permanently altered or that such cells are capable of continued abnormal growth in the absence of the pathogen.

At this point in the discussion we have reached the borderline between the self-limiting growths and the true or non-self-limiting tumors.

B. Non-Self-Limiting Tumors

Tumors of a non-self-limiting and trasnplantable type may develop in many plant species just as they do in many kinds of animals. In diseases of this type the new growth or tumor is composed of altered, randomly proliferating cells that reproduce true to type. The nature of this acquired autonomy is fundamental and constitutes the ultimate basis of the tumor problem.

Since the tumor problem is basically a cellular problem and since the fundamental similarity of cells and cellular processes in plants and animals is commonly recognized, certain plant tumors have for many years provided interesting and unique experimental material for studying the metabolic processes that underlie the tumorous state. Certain of the early workers, particularly C.O. Jensen, who is generally considered to be the father of modern experimental cancer research, and Erwin F. Smith, studied on a comparative basis the crown gall disease of plants and malignant animal tumors and they found that these two types of growth have much in common. In comparing plant and animal tumors it must be remembered, however, that there are certain developmental and functional characteristics commonly used in the differentiation of animal tumors that are more or less restricted to animals and cannot, therefore be carried over and applied to plant tumors. These have been dealt with in detail by White and Braun (1942), and Black (1949) and will not be considered further here. The most essential characteristic of being able to grow independently of any morphogenetic restraint, upon which all other diagnostic features must ultimately depend, is however, equally capable of expression in neoplasia of all higher organisms since it is a characteristic property of the cell itself.

While it is generally true that the more malignant a tumor cell is the greater is its capacity for growth, it is equally true that a tumor cell is not characterized primarily by its rate of growth. Certain normal cell types may grow and divide at far faster rates than do most tumor cells. Regenerating liver cells, for example, grow much more rapidly than do hepatoma cells present in that organ. Similarly, the meristematic cells at the apex of a rapidly developing root or shoot grow and divide at considerably faster rates than do plant tumor cells. Thus, it is not rapid growth but autonomous growth that characterizes the tumor cell.

Autonomy of neoplastic growth is not, however, a fixed and unvarying character but it has many gradations. At the one extreme are found the show-growing benign tumors that remain localized in the host. At the other extreme are the most malignant cancers that grow rapidly, infiltrate neighboring tissues, metastasize, and kill. Theoretically, autonomy of tumor cells requires within them something newly activated and distinctive, something that urges such cells to continued abnormal and unregulated growth. One of the difficulties encountered in attempting to define the nature of autonomous growth has resulted from the fact that some of the most diverse agencies known to biologists are capable of accomplishing this condition. Such factors as radiant energy, irritation, carcinogenic chemicals, microorganisms, and viruses have all been shown to be more or less effective tumor producers in certain animals and plants. The effectiveness of these agencies in eliciting tumor formation appears to be determined in large part by the hereditary constitution of the host. These agencies, with the exception of certain of the viruses, are concerned only with the inception of the tumors and play no role in the continued unregulated growth of a tumor cell. It is therefore necessary to distinguish between the proximate cause or causes concerned in tumor initiation and the continuing cause or causes responsible for tumor development.

Three non-self-limiting diseases of plants, each of which has a different and quite distinct initiating cause, have been studied fairly extensively. These are: (1) the crown gall disease in which an as yet uncharacterized tumor-inducing principle elaborated by a specific bacterium regularly converts normal plant cells to tumor cells, (2) Black's wound tumor disease which is of known viral etiology, and (3) Kostoff's tumors which have a genetical basis and which commonly arise as a result of such a nonspecific stimulus as irritation in certain interspecific hybrids within the genus *Nicotiana*. There are other true tumorous diseases of plants that have not as yet been extensively studied. Most interesting among these is White's spruce tumor, the etiology of which has not as yet been established.

Although the crown gall, wound tumor, and the genetic tumors have quite different and distinct initiating causes, there is every indication that the ultimate continuing cause is similar at a physiological level in the three types of growth.

The development of an acceptable concept designed to explain the continued abnormal growth of a tumor cell in the absence of any recognizable infective agent (except in the case of the virus-induced tumors) represents a very real challenge to students of abnormal growth. Such a concept must not only explain the underlying basis for autonomous growth but it must also account for the morphological, histological, and cytological peculiarities that characterize the tumorous state.

Plant tumors may assume a variety of growth patterns. These range from slowly growing benign to rapidly growing malignant, and from completely unorganized to highly organized teratomatous growths. All may be produced in certain test systems under precisely defined experimental conditions and all have their counterparts in animal pathology.

Growth in all animals and plants results either from cell enlargement or from the combined processes of cell division and cell enlargement. These two fundamental growth processes appear to be dependent for their development, in plant cells at least, upon specific substances that may be synthesized by, but are precisely regulated in, all normal plant cells. By varying the ratio of a factor limiting for cell enlargement and one limiting for cell division it is possible to obtain, with the use of certain normal cell types as a test object, either (1) a high degree of organization involving the production of numerous shoots, (2) a completely unorganized callus composed of hypertrophied and hyperplastic cells, or (3) the extensive formation of roots. The effect of certain limiting factors on the growth of plant cells and on the development of plant organs is clearly evident from studies of this type.

1. *Crown Gall.* The crown gall disease has been studied over the years more intensively than have other plant neoplasms. It has served as the experimental model in the field of plant oncology. Crown gall, which is initiated by a specific bacterium, affects plants belonging to at least 142 genera present in 61 widely separated botanical families. No monocot, however, shows unequivocal response to infection. The bacterium causing this disease possesses the ability to transform normal plant cells to tumor cells in short periods of time. Once the cellular alteration has been

completed, the continued abnormal growth of the tumor cell becomes entirely independent of the causal bacteria. The cells of such a tumor continue to grow autonomously at the expense of the host and under favorable conditions the resulting growths may reach enormous size. Crown gall tumors weighing up to 100 lb. have been described in the literature. In certain hosts such as sunflower and Paris daisy there are produced, in addition to primary tumors, secondary tumors that develop at points distant from the seat of the primary inoculation. The secondary tumors are of interest because they are commonly free of the bacteria that initiate the primary growth. This finding permitted the unequivocal demonstration of the truly independent nature of the crown gall tumor cell. Tumor cells isolated from such sterile tumors and planted on a suitable culture medium grew profusely and indefinitely on that medium. While the sterile sunflower tumor tissues increased in theoretical volume one hundred million million times in the course of one year on that medium, normal cells of the type from which the tumor cells were derived increased in volume only about two hundred and fifty times. This indicates, of course, that a profound change in the metabolism of the affected cells had resulted. Small fragments of sterile tumor tissue when implanted into a healthy host developed again into typical crown gall tumors that were similar in every respect to those originally initiated by the bacteria except that the new growths were free of the inciting organisms. Normal tissue implants, on the other hand, fused with the host and soon fell into the growth pattern of the host without ever forming tumorous overgrowths. Subsequent studies demonstrated that sterile tissue isolated from primary tumors of many different plant species showed the same type of growth autonomy in the host and in culture as did those isolated from the secondary growths.

The commonly found type of crown gall tumor is characterized both in the host and in culture by the mostly undifferentiated and completely unorganized growth of its cells. In certain plant species, the cells of which possess highly developed regenerative capacities, there may however be produced, in place of the characteristic crown gall tumor, a complex overgrowth or teratoma which is composed in large part of highly abnormal leaves and buds that show varying degrees of morphological development. Sterile tissues isolated from such organized but abnormal structures grow profusely in culture, as do typical crown gall tumor cells, on a medium that does not support the continued growth of normal callus tissue. Teratoma-derived cells differ from the commonly found crown gall tumor cells, however, in that they retain indefinitely, in culture, a highly developed capacity for organization. The surfaces of such cultures are covered with small, organized structures many of which appear morphologically to be leaves. Histologically, however, these structures are not usually made up of the well differentiated cell types of which leaves are normally composed. An attempt has been made to analyze more precisely the conditions which determine tumor morphology in crown gall. Of importance were (1) the strain of the bacteria used to transform normal cells to tumor cells, (2) the relative position that the altered cells occupy in a host, and (3) the inherent competence for regeneration prossessed by the affected cells. It was also concluded from this study that the tumor-inducing principle elaborated by highly virulent strains of the bacteria completely overwhelms the cellular factors concerned with differentiation and organization of pluripotent plant cells, while the principle associated with a moderately virulent strain is incapable of completely suppressing those factors. In the latter instance, the pluripotent teratoma cells retain, despite their conversion to tumor cells, highly developed regenerative capacities. The teratoma tissue is of particular interest because of its usefulness in studying the problem of recovery of crown gall tumor cells.

It is now generally believed that all living nucleated cells of a plant—except perhaps those that are heavily lignified—can be transformed to crown gall tumor cells. Experimental studies have shown, however, that such cells must be conditioned or rendered susceptible to transformation as a result of the stimulus accompanying a wound. It is only during a limited period in the normal wound-healing cycle that normal plant cells can be converted to tumor cells. Following wounding, the host cells gradually become susceptible, reaching a maximum vulnerability between the second and third days after a wound is made. Thereafter, the cells again become progressively more resistant as would healing progresses toward completion. When a wound is made, the cells in the region of the wound are activated and some two to three days later they begin to divide to heal the wound. Under normal circumstances, upon repair of the wound, the cells which participated in the wound response return again to a quiescent state. In the case of crown gall this return to normalcy is blocked. Some as yet undefined morphogenetic restraint is no longer applied to, or if applied is no longer effective upon, the tumor cells of the regenerating tissues. As a result, the altered cells continue their proliferation in an unregulated and autonomous manner. This unregulated growth results in disorganization of the tissue, in hypertrophy and hyperplasia of the cells, as well as in cytological abnormalities frequently found associated with the tumorous state in crown gall. It is clear, therefore, that the cells proliferating in the tumor are no longer callus cells for they no longer behave as callus cells do. They represent new cell types that have acquired as a result of bacterial action, new properties the most important of which is the capacity for continued unregulated growth in the absence of any recognizable infective agent.

Any attempt to account in physiological and biochemical terms for the abnormal behaviour of a tumor cell must, if it is to receive serious consideration, also account for the morphological, histological, and cytological peculiarities that characterize the tumorous state. Such a concept has been developed and the evidence upon which it is based is outlined briefly below.

Since tumor cells generally appear somehow to have acquired a capacity for autonomous growth as a result of unregulated growth and division, the problem of autonomy has been explored by means of an analysis of the factors concerned specifically with growth accompanied by cell division. As indicated above, the fundamental growth processes of cell enlargement and cell division appear to be dependent for their development in plant cells upon specific substances that may be synthesized by such cells. Crown gall tumor tissue has been shown, moreover, to be a rich source of both a factor normally limiting for cell enlargement and one limiting for cell division. As a result of recent work, it is now possible to delimit these two growth processes under fully controlled experimental conditions, with the use of certain plant cell types as test objects. When, for example, tobacco pith parenchymal cells are treated with synthetic growth substances of the auxin type such as naphthalene acetic acid, the pith cells enlarge greatly in size but they do not divide. It is only when a second growth factor such as 6-furfurylaminopurine—or the naturally occurring equivalent of that substance—is supplied to the pith parenchymal cells in addition to an auxin, that a profuse growth accompanied by cell division results. The 6-furfurylaminopurine without an auxin is ineffective in encouraging either an enlargement or a division of the pith cells. These findings demonstrate that two growth substances, one of which is concerned with cell enlargement and the other with cell division, act synergistically to promote growth and cell division in tobacco pith parenchymal cells. Normal tobacco pith cells do not and cannot themselves synthesize these two growth substances for, if they did, they would respond in the characteristic manner indicated above. Since the cellular systems responsible for the synthesis of these two growth substances appear to be solidly blocked in normal tobacco pith cells, a study was undertaken to learn how such cells would respond when transformed to crown gall tumor cells. The results of these studies, which are reported in detail elsewhere, demonstrated that when healing pith cells are converted to crown gall tumor cells, typical crown gall tumors develop. This simple experiment demonstrates that, although normal tobacco pith cells are not and cannot synthesize physiologically effective concentrations of either a cell enlargement or a cell division factor prior to their conversion to tumor cells, both substances are synthesized in greater than regulatory amounts following alteration. If this were not true, continued growth accompanied by cell division and, hence, tumor formation would not have resulted in the test system used in these experiments. It is clear, therefore, that an essential difference between a normal tobacco pith cell and a crown gall tumor cell appears to be concerned at a physiological level with the permanent activation of two growth-substances-synthesizing systems the products of which are specifically concerned with growth accompanied by cell division. The continued production is greater than regulatory amounts of the cell enlargement and cell division factors by the tumor cell could account for the continued abnormal proliferation of such a cell. Subsequent studies have shown, however, that additional metabolic systems are permanently activated during the transition from a normal cell to a fully altered, rapidly growing type of crown gall tumor cell. It has been found that alteration of normal cells to tumor cells is a gradual but progressive process. When, for example, the tumor-inducing principle responsible for inception of the crown gall tumor is allowed to act on plant cells for only 34–36 hours before being inactivated by thermal treatment, small, very slowly growing benign growth are elicited in a host. A 50-hour exposure of cells to the action of that principle results in tumors that grow at a moderately fast rate. If the tumor-inducing principle is allowed to act for 72–96 hours before being destroyed by heat, rapidly growing (potentially malignant) tumors result. It is also possible to obtain tumors showing varying degrees of neoplastic change by allowing the tumor-inducing principle associated with slightly virulent or moderately virulent strains of the crown gall bacteria to act on host cells throughout a 4- or 5-day period. Sterile tissue isolated from the three types of tumors described above and planted on White's basic culture medium retain indefinitely their characteristic growth patterns. Since these three types of tumors are derived from the same plant species, they are admirably suited for a study of the factors required for rapid autonomous growth. In these studies, the rapidly growing fully altered tumor cell was used as the standard. This cell type can synthesize in optimal or near optimal amounts all of the growth factors needed for its continued rapid abnormal growth from mineral salts and sucrose present in White's basic culture medium. The moderately fast growing tumor cell required that the basic medium be supplemented with glutamine, meso-inositol, and a cell enlargement factor (naphthalene acetic acid) to achieve a growth rate comparable to that of the fully altered rapidly growing type of tumor cell. The very slowly growing benign tumor cells altered in a 34-hour period required, in addition to the three compounds described above, asparagine as well as cytidylic and guanylic acids to achieve full growth. It is clear from experiments of this type that, as the crown gall tumor cell becomes more autonomous, its requirements in terms of externally supplied growth factors become less exacting. These studies demonstrate clearly, moreover, that a series of well defined growth-substance-synthesizing systems become gradually activated during the transition from the normal cell to the fully altered tumor cell, and the degree of activation of these systems determines the rate of growth of the tumor cell.

Normal cells of the type from which the tumor cells are derived do not grow on the basic medium. Thus, although the difference between the three types of tumor cells is quantitative since all grow continuously, though at different rates, on the basic medium, the difference between the tumor cells and the normal cell is qualitative. One quantative difference found to exist in these studies is the absolute requirement of the normal cell for 6-furfurylaminopurine or the naturally occurring equivalent of that substance. The addition of that compound to the basic media or to the supplemented culture media did not stimulate growth of any of the tumor tissues. The normal cells also possess, in contrast with the tumor cells, an absolute requirement as a supplement for a cell enlargement factor such as naphthalene acetic acid. The addition of 6-furfurylaminopurine and naphthalene acetic acid to the basic medium permit the very slow growth of normal cells. However, only if the basic medium is supplemented with glutamine, asparagine, inositol, guanylic and cytidylic acids, in addition to the auxin and 6-furfurylaminopurine, do the normal cells achieve a growth rate comparable to that of the fully altered, rapidly growing type of tumor cell. It thus appears that, as a result of the transition from a normal cell to a fully altered, rapidly growing crown gall tumor cell, a series of quite distinct but well defined growth-substance-synthesizing systems become progressively activated. This leads to the production by the affected cell of greater than regulatory amounts of these growth-promoting substances. The continued production in greater than regulatory amounts of these substances by the tumor cell could and most probably does account for the continued unregulated proliferation of such a cell. Precisely how the tumor-inducing principle associated with this disease accomplishes the simultaneous unblocking of several apparently distinct and quite unrelated metabolic systems remains unanswered. These results are nevertheless understandable if it is assumed that some as yet uncharacterized master reaction within the cell is specifically but gradually unblocked by the tumour-inducing principle and which, once activated, not only accomplishes the unblocking of several other growth-substance-synthesizing systems but also determines the rate at which the entire series of metabolic events concerned with growth and cell division proceeds.

The concept of growth autonomy presented above finds additional support in other directions. It has been possible to reproduce, under precisely defined experimental conditions and with the use of certain normal cell types as a test object, not only the morphological growth patterns (slow and rapid disorganized growths, teratoma-like structures) but also the histological (hypertrophy and hyperplasia leading to disorganization and loss of function) as well as the cytological (multinucleate giant cells, etc.) abnormalities that characterize the tumorous state in crown gall. This was accomplished by varying the proportions of the cell enlargement factor and the factor limiting for cell division in an otherwise suitable culture medium on which the normal cells were planted. These artificially stimulated normal cells, in contrast to crown gall tumor cells, are self-limiting growths and when the externally supplied stimuli are removed, their growth promptly stops. The fact that such stimulated normal cells commonly show histological and cytological characteristics of true tumor cells but the themselves self-limiting growths indicates that the observed cellular abnormalities are the result rather than the cause of the tumorous state.

Quite possibly, a cell acquires the capacity for autonomous growth as a result of the permanent activation of a series of growth-substance synthesizing systems the products of which are concerned specifically with growth accompanied by cell division. These systems are precisely regulated in all normal plant cells.

Here, then, is a rare example of a disease in which an infective agent in some as yet uncharacterized manner regularly converts normal plant cells into new cell types in short periods of time. Once the cellular transformation has been accomplished, the pathogen no longer plays a role in the continued development of the disease. The altered host cells become pathogenic by virtue of the fact that those cells have acquired a capacity to synthesize greater than regulatory amounts of growth-promoting substances of a type that are limiting for growth and cell division in normal plant cells.

Although growth-promoting substances elaborated by the tumor cells themselves appear to be responsible for the continued unregulated growth of such cells, other sources of growth hormone may influence the morphological growth patterns as well as the rates of growth of the resulting tumors. This is particularly evident when dealing with crown gall tumor cells that grow slowly and possess relatively low grades of neoplastic change. For example, the tumor-inducing principle elaborated by certain attenuated strains of the crown gall bacteria initiates very slowly growing tumors on plants such as the tomato. The application of synthetic hormones of the auxin type to such normally slowly growing tumors results in the formation of large rapidly growing tumors. In this instance, the effect of the growth-promoting substance on the cell is only temporary and, although growth is enhanced, the increased growth rate does not become an intrinsic property of the cell itself. Not only the rate of growth but also the tendency of tumor tissue to organize may be hormonally influenced. When, for example, the tumor-inducing principle elaborated by a moderately virulent strain of the bacterial pathogen transforms pluripotent cells at the cut stem tip of a tobacco plant, complex tumors or teratomata arise. When, on the other hand, this same tumor-inducing principle alters similar cells in an internode of a plant containing a functional apical bud, and hence hormone-producing center, typical unorganized crown gall tumors develop. In this instance, the functional apical bud can be wholly replaced in suppressing the tendency of the tumor tissue to organize by synthetic growth hormones. The effect of the hormone in suppressing organization in the tumor tissues is only temporary, as evidenced by the fact that tissue isolated from either the organized teratomata or unorganized tumors, when planted on a culture medium, shows identical growth patterns. In both instances, the tissues are characterized by a capacity to organize highly abnormal shoots and buds. It can, therefore, be concluded that the greater the degree of the primary change in a tumor cell and, hence, the greater the growth-substance output by that cell, the less effective are externally supplied hormones in stimulating growth. The growth of fully altered, rapidly growing crown gall tumor cells is, in fact, inhibited when such cells are treated with growth substances of the auxin type.

2. Wound Tumor

A second plant disease which fulfills all of the criteria of a true neoplastic disease is one caused by a typical virus. This malady is known as the wound tumor disease. The etiological entity responsible for this disease is insect-transmitted and is one of a group of viruses that multiply in the insect as well as in the plant host.

On a large variety of plants, the symptoms produced in response to the virus involve morphogenetic disturbances. A survey of potential hosts indicated that at least 43 species belonging to 20 plant families produce symptoms typical of the disease including irregular enlargements of the veins in leaves and tumors on the roots. Other symptoms observed included leaf curling and distortion, leafy outgrowths from the undersides of the veins, vein tumors, distortion of petioles, and shortening of internodes with a resulting dwarfing of the infected plant.

The variations in symptoms vary from barely detectable to those that are extremely pronounced. In *Portulaca* the tops of the infected plants bear no symptoms, but the roots contain many small tumors. In some plants such as sorrel and sweet clover the response to infection involves the production of large tumors which in the case of sweet clover appear on the stem. Investigations on the latter host demonstrated that the hereditary constitution of the plant affects the number, size, distribution, and morphology of the tumors. In some strains the root tumors may be so large and numerous that they fuse together, while in others so inconspicuous that they may easily be overlooked. Similarly, in some clones, stem reaction involves the formation of many large tumors; in others they are rarely found. Moreover, hereditary influences on the stem and root may act independently. It has been found, for example, that one clone produces large tumors on stems and roots while another produces many large tumors on the roots but only occasional tumors on the stem.

That the genetic basis which causes a clone to be highly predisposed to tumor formation involves not merely a susceptibility to the virus but also an inherent tendency toward tumoruous proliferation is indicated by the findings of Littau and Black (1952). These investigators found that the inbred B21 clone of sweet clover, which responds actively with tumor formation as a result of virus infection, produced five spontaneous tumors over a period of several years. That these new growths were not the result of accidental virus infection was adequately demonstrated. The authors have compared their B21 line of sweet clover with strain C_3Hb of mice which have lost the mammary carcinoma virus and yet show a strong inherent tendency toward the development of mammary carcinoma.

In this, as in other neoplastic diseases of plants, wounding plays an important role in the initiation of the disease. Tumors develop from accidental or artificially made wounds, from points of stress, as well as in those regions where lateral roots emerge. Black and Lee (1957) have also demonstrated that when infected plants are treated with an auxin, a pronounced increase in the number and size of stem tumors results. Thus, at least three factors appear to be essential for the initiation and development of tumors in this disease: (1) the inciting virus, (2) a wound or some similar stimulus, and (3) the hereditary constitution of the host.

The function of the wound is not yet clear. Brakke et. al. (1954) have suggested that perhaps the response of a cell to a wound permits an increase in virus concentration and that the high concentration maintains the cells in a state of active proliferation. The cells present in tumors of sweet clover have been found to contain about one hundred times more virus than do nontumorous portions of a diseased plant. Another possibility presents itself, however. It may be that virus-infected and hence potential tumor cells do not develop into a neoplastic growth until they are first stimulated to divide by noncarcinogenic processes such as wounding, application of hormones, etc. Once cell division is initiated as a result of a nonspecific stimulus, the infected cells are no longer subject to the morphogenetic restraints in a host which normally return the cell to quiescence after wound healing has gone to completion. Kelly and Black (1949) have raised the very pertinent question as to why one cell in the pericycle of an infected sweet clover root develops into an organized lateral root, while other pericycle cells in close proximity develop into disorganized tumors. There are several possible answers to this question. Since lateral root formation occurs earlier than tumor formation, it may be that the more embryonic tissue has such tight restraints placed upon it that the infection does not lead to its disorganization. On the other hand, lateral roots, initially at least, appear to involve the xylary pericycle while the tumors generally involve the phloic pericycle. These differences may reflect divergencies in morphogenetic response at the two sites. A third possibility rests on the assumption that only certain cell types are infected by the virus. The leaf hopper-transmitted viruses, of which the wound tumor virus is a representative, appear to show a predilection for phloem tissue. Since lateral root formation initially involves the xylary pericycle, it may be that at the stage or root development that permits lateral root initiation, only that portion of the pericycle closest to the of phloem is infected with virus. It would, therefore, not be until lateral root formation involved tissue close to the phloem that a tumor would develop, because not until then would the cell division stimulus associated with lateral root formation involve virus-infected cells.

Once tumors are initiated, the virus remains closely associated with the tumorous tissue. These tumors possess a capacity for indefinite growth, both in the host and in culture. Sterile tumor tissue isolated from sorrel and planted on White's basic culture medium initially grew rather slowly and doubled its volume every 3 weeks. Subsequently, Burkholder and Nickell (1949) devised a more suitable culture medium which permitted far more rapid growth of the tumor tissue. The tumor tissue in culture was found to possess an unusually high requirement for phosphorus. This was interpreted to reflect the need of the multiplying virus for this substance. Similarly, an RNA hydrolyzate, and more specifically uracil, was found to exert a stimulatory effect on the growth of the tumor tissues.

There is no question about the fact that specific viruses are etiologically implicated in the wound tumor disease of plants as well as in certain virus-induced tumors of animals. The question that arises, however, is: How does the virus induce the neoplastic state in a host cell? Viruses cause cells of plants and animals to respond to infection in many different ways. Few elicit tumorous growths. Even the tumor-inducing viruses may infect cells and not exert a stimulatory influence on such cells. What is it then that causes certain viruses to induce cells to proliferate in an essentially unregulated and uncontrolled manner? There is a partial answer at least in the case of Black's wound tumor disease. The virus-infected tissue, like crown gall tumor tissue, acquires a capacity to synthesize growth-promoting substances concerned specifically with growth and cell division. This is evidenced by the fact that the tumor cells grow rapidly and indefinitely on a culture medium that is lacking in certain growth factors needed for the continued growth of normal cells. It thus appears that the presence of the virus confers upon the cell the ability to produce such growth-promoting substances in greater than regulatory amounts and as a result the cell achieves a physiological autonomy.

3. *Genetic Tumors*. The formation of non-self-limiting tumors, in the development of which no demonstrable infectious agent is involved, is a regular occurrence in certain plants. In these instances, it is the genetic

constitution of the cells within a plant that appears to be of importance. The most thoroughly investigated examples of this type involve the development of spontaneously occurring tumors in a large number of *Nicotiana* hybrids. Kostoff (1930) demonstrated that when appropriate *Nicotiana* species are crossed, tumors develop in all hybrid offspring. These hybrid plants are interesting because they are, for the most part, perfectly organized both morphologically and histologically during their period of active growth ceases, a profusion of tumors may develop on all parts of the plants. It is, however, possible, according to Kunkel (1954), to initiate a tumor at almost any period in the development of a plant by making a wound in the vicinity of a vascular bundle. The cells of such plants behave as do normal cells until they are stimulated to divide. Once the cells of this plant are activated as a result either of natural processes or artificially induced irritation, they no longer respond to the morphogenetic restraints that return a normal cell to quiescence. The genetic constitution of the cell, in this instance, is critical. Only such a nonspecific stimulus, such as irritation, is required to transform the potential tumor cells, of which the hybrid plant is composed, into actively proliferating autonomous plant-cell types.

Since Kostoff's original description of this phenomenon, the number of tumor-producing combinations has increased considerably. The tumors apparently arise in tissue that has been stimulated to division by natural or artificial processes and, once initiated, the tumors retain the capacity for unlimited disorganized growth both *in vitro* and *in situ*. Normally, tumors on the stem develop at the site of leaf or petal abscission. A number of hybrid combinations have been found to produce tumors on the roots only. In those many instances in which tumors develop on all parts of the plant, tumor formation on the root often precedes the formation of overgrowths on the shoot. This increased response on the part of the root may result for several reasons. Lateral root formation may represent a stimulus similar to wounding and may, therefore, play a role in tumor initiation similar to that found in the virus tumors. Irradiation of hybrid plants hastens the onset of tumor formation and increases significantly the number of tumors that develop. Chemical irritation may also initiate tumors. Kehr and Smith (1954) reported that leaves of certain hybrid combinations accidentally sprayed with a mixture of turpentine, whiting, and white lead, produced tumors at almost every spot where droplets of the spray mixture fell on a leaf.

The genetic tumors appear to represent a lower grade of neoplastic change than do most crown gall tumors. This is evidenced both by the fact that the stem tumors show a strong tendency to organize abnormal structures and that they seldom, if ever, reach the size and state of disorganization of crown gall tumors initiated by highly virulent bacteria. Although these tumors are almost never directly fatal to the plant, they do represent a considerable burden to the plant.

Kehr and Smith (1954) attempted to analyze the precise genetic basis of these tumors by breeding a considerable number of diploid and polyploid combinations of *Nicotiana glauca* × *N. langsdorffii*. From the data obtained it was concluded that the tumor-forming nature of the hybrid remains relatively unchanged regardless of the ratio of *N. glauca* and *N. langsdorffii* genomes as long as at least one complete genome of each species is present in the hybrid. It was found, further, that, although spontaneous genetic tumors develop when all twelve *N. glauca* chromosomes are combined with a diploid complement of *N. langsdorffii* chromosomes, no tumors developed in hybrid plants when only one or a few *N. glauca* chromosomes are present in addition to the diploid *N. langsdorffii* genome.

Recently Näf (1958) has approached the problem from a different point of view. This worker has divided all of the parents of tumorous hybrids into two groups which he arbitrarily designated as "plus" and "minus." Näf found that if an intragroup cross is made either between two "plus" species or two "minus" species, the offspring never develops tumors. On the other hand, crosses made between a "plus" species and a "minus" species produce tumorous offspring. Of a total of more than 50 crosses tested; very few exceptions to the above rule were found. Näf explains these exceptions on the basis of relative "plusness" and "minusness," similar to Hartmann's concept of relative sexuality. It was concluded from these studies that the critical contributions to tumor formation of the "minus" parents differ from those of the "plus" parents. These contributions, since they are of a genetical nature, must be reflected in parental metabolism and it should be possible to characterize them on a physiological level.

As in the case of crown gall and Black's virus tumor, tissue isolated from the hybrid tumors and planted in culture is capable of synthesizing all of the growth factors required for its continued abnormal growth from mineral salts and sucrose present in a basic culture medium. What is particularly interesting in the case of the hybrid is that cells obtained from nontumorous portions of hybrid stems also become autonomous upon isolation and culture. Such tissue is truly tumorous and, as White

(1944) has shown, it can be grafted to one of the parents, *N. glauca*, where the implants again develop into tumorous overgrowths. Hybrid tumors tissue does not commonly differentiate and organize when grown on a semisolid medium. White (1939) found, however, that when such tissue is immersed in a liquid medium, it tends to organize shoots and leaves. It was presumed that oxygen gradients influenced the differentiation process. Skoog (1944) found, however, that the tendency to form buds and leaves can be completely suppressed by the addition of 0.2 p.p.m. of indoleacetic acid. Low concentrations of indoleacetic acid were found not only to suppress organization but also to stimulate growth. The indoleacetic acid effect was reversible by raising the level of certain nutrients such as sucrose, KH_2PO_4, and $Fe_2(SO_4)_3$. Subsequent studies on tobacco demonstrated that bud formation on isolated stem segments of tobacco depends upon an adenine: indoleacetic acid ratio. Adenine favors bud development which, in turn, may be inhibited by the addition of indoleacetic acid. More recently, Skoog and Miller (1957) have reported that 6-furfurylaminopurine is far more effective than adenine, in this respect.

The capacity of hybrid tumor tissue to organize shoots and buds under certain conditions does not reflect a tendency of such tissue to return to normalcy since these organized structures are composed entirely of potential tumor cells and they would appear, therefore, to be organized tumors.

The following picture emerges from studied on the non-self-limiting neoplastic diseases of plants. It is apparent from this discussion that several quite distinct agencies can bring about the tumorous state. The biological aspects of this state are characterized by growth autonomy. This phenomenon can be adequately explained on the basis of the ability of the tumor cell to synthesize certain growth-promoting substances in greater than regulatory amounts. The rate of growth of a tumor cell appears, moreover, to be a function of the degree to which the growth-substance-synthesizing systems within a cell are activated. The ultimate mechanism by which the growth-substance-synthesizing systems become activated in plant tumor cells is unknown. The production of these substances by the cell is, however, presumably enzymatic in nature.

Since enzymatic reactions are commonly considered to be gene controlled processes, it might appear that the normal gene complement is somehow modified in the plant tumor cell. This could conceivably be accomplished by somatic mutation at the genic level. There are several reasons for questioning somatic mutation as being the cause of the physiological autonomy that underlies the tumorous state in plants. The first of these is that a specific virus has been implicated etiologically in one of the diseases in question. It might be argued, however, that the virus induces a mutation at the genic level in cells and that once this change is effected, the virus is no longer needed for the continued abnormal growth of the tumor cell. While this possibility cannot at present be ruled out in the case of Black's virus tumor, other closely related viruses that produce abnormal growth patterns in plants have been eliminated from their hosts by thermal treatment with a resulting complete recovery of such plants. Plant viruses of this type do not appear, therefore, to induce permanent modifications in plant cells. The continued abnormal behaviour of the cell seems to be dependent upon the continued presence of the virus. In this respect, the virus-induced tumors behave as does the clubroot disease of the crucifers in which continued cell stimulation is dependent upon the continued presence of the pathogen.

The reported recovery of crown gall tumor cells also appears to argue against somatic mutation as the cause of the physiological autonomy found in this type of tumor cell. By forcing abnormal tumor buds present in crown gall teratomata into very rapid growth by a series of graftings to healthy plants, a gradual but ultimately complete recovery of such cells was achieved. These findings suggest that crown gall tumor cells may recover if they are forced to divide with unusual rapidity at the stem apex. They suggest further that the factor responsible for the continued abnormal proliferation of the crown gall tumor cell is an autonomous or partially autonomous entity that is subject to the effects of dilution in cells that are forced to divide with great rapidity.

The findings reported above are very suggestive of some encountered in microbial genetics. Studies such as those presented by Ephrussi (1951), Sonneborn (1946), and Spiegelman (1954) indicate that certain self-duplication cytoplasmic factors as well as the nuclear genes may serve as determinants of hereditary differences in a cell. Certain of these cytoplasmic entities appear, moreover, to be concerned with enzyme production. Therefore, mechanisms quite different from somatic mutation at the gene level can be postulated to explain the continuity of tumorous properties from one cell generation to the next. It may well be that it is in the particulate cytoplasmic fraction of the cell rather than in the nucleus that changes occur which account for the physiological autonomy that underlies the tumorous state in plants.

5

EFFECT ON REPRODUCTION

Flowering plants must attain a minimum size or age before they can reproduce. During the vegetative stages of growth and development, a plant is especially dependent upon nutrients. Energy is required for the transition from vegetative to reproductive phases and for the process of reproduction itself. Reserve foods supply this energy, and the plant, accordingly, must contain a sufficient reserve of foods to carry out the reproductive process. Reproductive organs, however, do not usually manufacture foods to any appreciable extent and seeds in particular are heterotrophic. In many species, especially those with an indeterminate growth habit, different parts of the same plant may simultaneously carry on vegetative and reproductive functions.

Light and temperature directly affect the initiation and development of floral primordia and the reproductive organs that develop later. Leaves, apical buds, and other vegetative organs are the receptors of these stimuli. Thus, the environment may determine whether anthesis begins, is retarded, or is suppressed. For example, sugar beets may flower prematurely in one environment or remain vegetative indefinitely in a humid environment. The course of reproduction may embrace more than a season, and, in biennials and woody plants, more than one year. Therefore, the plant is ordinarily subjected to wide variations in climate and nutrition during its life, factors that are critical to the over-all reproductive process.

Reproduction proceeds through stages of ripeness-to-flower, floral initiation, anthesis, fruit set, and maturation of fruit and seeds. As reproductive activity begins, the physiology of the plant is altered. Vegetative growth decreases and is continuously readjusted in accordance with the demands of reproduction. For a given plant and environment, the quantity of leaves, stems, and roots determines in a gross way the number of flowers formed and particularly the number of fruits that come to maturity. Especially in an unfavorable environment, there is a fluctuating competition between fruits and roots, leaves, or even other fruits on the same plant. As a consequence, reproduction slows down the vegetative growth but does not inhibit it entirely, and vegetative growth may proceed slower and then faster in succession in different stages of reproduction. The complexity of these relationships in such that the plant as a whole must usually be considered in a discussion of the pathological aspects of reproduction.

Seeds and pericarp make very different substrates for pathogens. The heterogeneous nature of reproductive organs is evident on caryological examination. In addition, during the maturation of fruits, especially fleshy ones, seeds become dehydrated and show a marked increase in nitrogen content, ad in condensation of reserve foods and finally appear isolated from the placenta, whereas in the pericarp, cell walls are hydrolyzed, simple sugars are formed, and water content increases.

Diseases of the pericarp have little influence on seeds unless there is premature fruit drop. Conversely, seeds may be killed by late frosts whereas the pericarp is uninjured. When seeds are physiologically mature and firm, lysigenous breakdown of the pericarp begins adjacent to them and is intimately dependent upon their maturity. On the other hand, the metabolic activities of the pericarp may produce physiological active substances such as ethylene, which at appropriate concentrations may play a role in inhibiting seed germination. Thus, there is a correlation but no parallelism in the growth of the various parts of the fruit.

I. PARTHENOCARPIC AND UNDERDEVELOPED FRUITS

Seedless and parthenocarpic fruits exemplify the correlation in development of the fruit parts. When parthenocarpy occurs naturally, it is assumed that the auxin content of the unfertilized ovule is sufficient to promote fruit development. The same results can be obtained by the application of auxin sprays to plants that do not undergo parthenocarpy naturally. These fruits are seedless, contain sterile pseudo-seeds, or, by partial reversal of the normal growth correlation, parthenogenetic embryos, *sensu lato*.

In fruits, the embryo, albumen, and pericarp are correlated in development. The embryo and albumen both arise through separate fertilizations on the part of the pollen nuclei. Therefore, in sterile seeds with an abnormal; nongerminative embryo, the endosperm may be quite normal, e.g., in *Ruta graveolens*. A similar situation occurs in Bartlett pears. In a favorable environment, the Bartlett pear matures fruits that arise from vegetative or pollen-induced parthenocarpy and are produced uniformly, abundantly, and of good shape. Unfortunately in other cases parthenocarpy is partially induced.

In the olive tree, very small, persistent pseudodrupes originate from pollinated flowers in which elongation of the pollen tubes ceases in stylar tissues. In these fruits, the useless pericarp is complete in all its parts, whereas the stone which consists of the hardened endocarp, contains pseudoseeds showing parenchymatous degeneration of the 4 ovules and no embryo. Factors that lead to arresting of pollent tube growth and therefore to the formation of pseudodrupes in olives are incompatibility factors, and unfavorable environmental conditions such as high humidity and low temperature.

Parasites may cause similar responses. Thus, olive pseudodrupes may be the consequence of attack by insects, sch as *Coccus oleae* and *Mytilococcus ulmi*. The same intrinsic or environmental causes are thought to be responsible for some types of "shot-berries" in other fruits, e.g., grapes.

II. NUTRITIVE DISTURBANCES

Nutritive diseases act on the whole plant. Consequently, they influence the reproductive organs through the vegetative ones. El Hinnawy (1956), who recently studied this problem, concluded that the influence of minerals on the flowering response of long and short day plants is indirect and is exerted through their effect on the production of auxin, and therefore on growth and the production of material for flower bud formation.

A. The Carbohydrate: Nitrogen Ratio

In past decades, there was a prevalent belief that the relation between the life of the individual and the preservation of the species was largely dependent upon the C : N ratio, the influence of which on flower initiation was expressed in Kleb's or Fischer's rule. All the conditions that either favor an accumulation of assimilates or hinder the absorption of water and nitrogen salts increase the ratio of carbohydrate to inorganic nitrogen and consequently favor the transition of the plant to the reproductive stage. On the other hand, all conditions which cause the opposite effects favor vegetative growth.

After the discovery of photoperiodism, the importance of C : N ratio was questioned and flower induction is now thought to be controlled by flowering hormones—the florigens. These have not as yet been isolated.

Floral initiation is not fundamentally dependent upon the C and N supplies. Rather the proper photoperiod and other environmental factors as well as nutrition are necessary conditions to stimulate the formation of florigens. In healthy plants, the transition from vegetative to reproductive stages requires a level of light and nutrition exceeding the minimum needed by vegetative growth. With the onset of reproduction, vegetative growth is reduced. N and particularly C accumulate in the plant, and as a result the C : N ratio increases. However, this accumulation seems a consequence of flower induction and not a cause of it.

Once floral primordia are initiated, they as well as the blossoms and fruits and strongly influenced by the C and N supplies and their broad interrelationships are very important. A satisfactory understanding of these complex phenomena has not been gained as yet; coordinate studies on growth promoting substances and on food competition throughout the plant are badly needed.

B. Nutrient Excesses and Deficiencies Affecting Reproduction

The above discussion of floral induction refers to plants living within "the limits of health"; if the abundance or the deficiency of one nutrient in relation to another exceeds a certain range, the plant enters into a real pathological condition.

The supply of carbon directly influences flowering through its presence in a carbohydrate precursor of florigens and through the influence of carbohydrate translocation on the movement of the flowering stimulus.

Flower induction can be effected with very short exposures to illumination and with light intensities well below the compensation point. Later, a large supply of carbohydrates is necessary for reproductive growth and, at least in some cases, it has been found that carbohydrates directly influence the development of normal reproductive organs.

The nitrogen nutrition of plants affects reproduction in a number of ways. Some plants seem to regulate nitrogen absorption regardless of the available supply, but the response of different species to nitrogen is unpredictable. The assumption that nitrogen deficiency will promote flowering in long-day plants and slightly delay it in short-day plant proved to be erroneous; Murneek's conclusion (1948) still seems valid that nitrogen is one of the most crucial nutrient elements in the initiation of reproduction. Nitrogen promotes growth in accordance with the principle that without nitrogen there is no growth hormone and consequently no growth. Therefore, an abundant nitrogen supply is usually required in the reproductive stages subsequent to floral induction. An excess or a deficiency of nitrogen can easily impair flowering or fruit set and the former condition frequently induces as increase in the number of members of flower verticillia, virescences, proliferations, and recrudescences of reproductive organs. Excessive nitrogen also increases the susceptibility of blossoms to low temperature. During maturation of fruits, excessive nitrogen may excite more complex responses on the part of the plant as in "Baldwin spot" of apples. Nitrogen deficiency may directly cause pistil abortion, affect pollen germination, and induce flower shedding and fruit drop. In dry climates, nitrogen deficiency is one of the most common causes of staminate flowers in the olive tree and nitrogen fertilization is, therefore, advocated so that otherwise symptomless trees will form normal flowers and set fruits. Biennial bearing in olives in dry climates is reduced by nitrogen dressings in April of the preceding year.

Although phosphorus does not affect flower initiation, according to El Hinnawy (1956), it accumulates in the developing primordia and flower buds. Phosphorus deficiency may lessen the fruit set and is a recognized factor affecting earliness of cereals and upland cotton, among other plants.

Phosphorus, however, does not seem highly important in some fruit crops. Thus, in peach, phosphorus can almost be considered a secondary nutrient, ranking below nitrogen, potassium, calcium, and magnesium. This behavior may be a result of the interaction of phosphorus with other elements, especially nitrogen.

Potassium plays an important role in carbohydrate metabolism; potassium deficiencies reduce set, size, sugar content, and color of fruits. An excess of available potassium, which is often related to high nitrogen levels, can indirectly damage reproduction because of its strong antagonism to calcium and, to a lesser extent, to magnesium, zinc, and other nutrients. Thus, an excess of potassium tends to accentuate the symptoms of blossom end-rot in tomatoes and of "Baldwin spot" in apples.

Boron deserves particular mention in regard to reproduction. In very low concentration (1 to 100 p.p.m.), it regulates the water absorption of certain pollens, prevents their bursting, and favors their germination. Also, it activates the development of the pollen tube. These functions of boron explains its presence in the nectar of the Nymphaeaceae and in the stigmatic secretions of tomato and other plants. The stimulation by boron of the elongation of pollen tubes is attributed to its influence on oxygen uptake and sugar absorption, but it may also involve the synthesis of pectic materials in the cell walls of the elongating pollen tube. In boron-deficient grapes, fruit set and yield are severely depressed, and fruits are seedless and undersized. Boron-deficient apples show internal and external corking. Preharvest drop and breakdown in late stored apples are also associated with the boron content.

A slight disturbance in boron nutrition sometimes affects a single stage of reproductive activity. Thus, an incipient boron deficiency in vigorously growing grapes and pears may cause failure of fruit set although there are no foliar symptoms. Sometimes this behaviour is attributed to a high boron requirement of the plant when blossoming. In other cases it is attributed to the unavailability of boron in heavy, water-logged soils during the spring, or to the dilution of boron induced by rapid development of vegetation following nitrogen fertilization.

According to Brennan and Shive (1948), the influence of calcium on carbohydrate translocation is a result of the relationship of calcium to boron. Recently, Joham (1957) has judged the influence of calcium on carbohydrate translocation to be similar to, but independent of, boron.

Calcium deficiency per se exerts other, more direct, influences on reproduction. Thus, in cotton it affects the earliness of flowering, the number of flowers, and the weight of bolls. "Baldwin spot" of apples, often attributed to an excessive development of foliage, has been associated with calcium deficiency in the fruit.

A zinc deficiency prevents the normal production of tryptophan, a precursor of indoleacetic acid, and results in an increase in blossom and fruit drop.

Recently, iron-deficient cocklebur has been studied during photoinduction treatment. Staminate flower primordia were delayed in appearance and developed more slowly. In pistillate inflorescences, mature burs developed abnormally or failed to be produced. The influence of iron during photoinduction was more pronounced than that of boron or magnesium.

Deficiency of magnesium seems especially effective in delaying flowering in such plants as mustard. It has also been repeatedly associated with fruit drop.

III. WATER AND REPRODUCTION

When water in the soil or in the atmosphere is in excess or is deficient, or when the amount of water available undergoes sudden fluctuations, damage is likely to result. Unfavorable water relations in the soil do not always produce the same effect as unfavorable moisture relations in the atmosphere. For instance, an excess of humidity around flowers may prevent the access of oxygen and reduce abscission, whereas an excess of water in the soil increases abscission.

Excessive moisture generally interferes with the transition of the plant to the reproductive stage. When the atmosphere is too humid, flowering is delayed, fertilization is reduced because insect movement is reduced, stigmatic secretions are diluted, and pollen grains are not dispersed by wind and burst. Consequently fruit set is curtailed.

High atmospheric humidity may affect reproduction, even in fully developed fruits and embryos. Wheat embryos may germinate in the inflorescences still in the field. Intumescences of pod valves and of bean and pea seed are also attributed to high humidity.

Drought conditions, even if short in duration, may have profound effects upon pollen formation and vitality. At meiosis, drought may cause diploid and tetraploid microspores and pollen grains in *Tradescantia*. If some wilting occurs, the pollen may fail to develop. During bloom, water inadequacy, therefore, reduces fruit set and causes blossom shedding.

Later on, a moderate moisture deficiency may increase firmness and keeping quality of apples and pears, but—especially in arid climates—it affects the leaf: root and leaf: fruit ratios and the competition among fruits. Sometimes, when fruits have attained their final size, competition between fruits and vegetative organs favors the fruit itself. Thus, in the San Marzano variety of tomato, the leaves—which have a thin cuticle—may wilt, whereas the fruits hold their moisture more efficiently by means of the mucilages and pectic substances they contain, and mature almost normally.

Usually, however, a water deficit causes a reduction in size, number, and quality of mature fruits. In mild cases, there is a partially reversible withdrawal of water from fruits, causing a shrinkage in olives, citrus fruits, etc. For this reason, late spring and summer irrigation increases the frequency of annual bearing in the olive tree, lessens fruit drop, and increases the size of individual drupes. In temperate fruit crops, such as apple and pear, the flower buds are laid down in the year preceding flowering. A mild water stress during the critical period may slow down vegetative growth and enhance flower initiation. Once induction has occurred, the subsequent stages of flower development require an adequate water supply.

In some tropical fruit trees, the dry season is the only rest period and the succession of dry and rainy periods is necessary for a normal periodical growth. In *Litchi chinensis* floral initiation and flowering occur satisfactorily in Canton (Southern China), where the above conditions are found, but floral initiation is rare in Hawaii and there recourse is made to the use of such auxins as sodium naphthalene acetate.

Drought can prevent after-ripening and induce "secondary dormancy" in seeds.

IV. TEMPERATURE AND REPRODUCTION

In plants requiring a thermoperiod to go into the reproductive stage, temperature determines the initiation of flowering. The apical bud is the receptor of the stimulus and in the absence of a favorable thermoperiod, such plants either do not form floral primordia or fail to flower. If a favorable stimulus is received out of season, it can induce plants to flower at that time, e.g., the preflowering of sugar beets when winters are intermittently mild.

A minimum daily range of temperature is often necessary for good fruit set. Diurnal thermoperiodicity equilibrates photosynthesis and condensation reactions in daylight and respiration and growth at night. For the tomato, night temperatures should be at least 6°C lower than daytime temperatures. The best temperatures for setting large crops of tomatoes in Texas, range from 13 to 20°C at night and from 21 to 30°C during the day. Daily excursions of temperature are also needed for the after-ripening of some seed and can act as a substitute for light on light-sensitive seed.

Excursions of temperature on an annual basis are sometimes required. Thus, the annual periodicity of growth in subtropical trees is often impaired by the absence of a moderate winter chilling. Inflorescence development in olive and subsequent fruit production are generally proportional to the amount of chilling received. For this reason, olive yields poorly in the African highlands.

High temperatures are a common cause of blossom shedding and fruit drop. The most sensitive parts of flowers are pollen and stigmas. Thus, tomato pollen is inactivated when temperatures exceed 32°C although the variety of the plant and its vigor affect this relation. Thus, the Hotset variety of tomato will set fruit at temperatures from 3 to 5°C higher than the maxima that are usually tolerated.

Although high temperatures break dormancy of some seeds, they induce secondary dormancy in others and temperatures above 30°C are not favorable to the germination of some species of seed.

The first effect of low temperatures may be shown by pollen. Temperatures below 15°C at the time of dispersal may devitalize maize pollen. Reproductive organs are highly susceptible to frost injury. According to Modlibowska (1956), the various parts of the flower are not equally susceptible. Petals may be harmed when male and female organs are still uninjured and vice versa. Often the filaments are damaged, whereas the gynoecium is still intact. Early autumn frost may harm the flesh of fruits that are low in sugar at maturity or that ripen late in the fall or during the winter. Thus, according to Azzi (1928), mature olives are injured by temperatures of – 0.4°C but even in these cases early frost causes little or no damage to reproduction.

Late spring frosts affect reproduction frequently. Modlibowska (1956) has shown that the economic consequences of late frost on the yield are less than is thought. Even when as many as 95% of the flowers are killed, the remaining ones produce a good crop if trees are well tended and pest and disease control is adequate. The effect of intense and repeated frosts on such early flowering trees as the almond may be so great, however, that the reproductive activities for the entire year may be suppressed.

The cold susceptibility of reproductive organs in different species varies widely. In apple and cherry the pistils are first injured at the base of the style. In pear the ovary is most susceptible. In cherry and plum the first lesions on the ovary are external. In apple and pear the centres of the flower and young fruit are injured earlier than the pericarp. Later the reverse is true. The susceptibility of flowers and fruits to frosts varies with the stage of growth, even in different varieties of a single species. Among apple varieties, "Belle de Boskoop" is equally sensitive at all stages; "Bramley's seedling" is most sensitive at the green button and pink button stages, but becomes hardier later; the reverse is true in "Ellison's orange."

The minimum temperature tolerated by flowers depends on their physiological condition, their nutrition, and the rapidity of growth. Thus rapid growth in a humid and warm environment predisposes flowers to frost injury.

Attack by pathogens may predispose flowers to frost injury. Thus, Modlibowska (1956) has shown that infection by *Stereum purpureum* increases susceptibility of plum flowers to low temperatures.

V. LIGHT AND REPRODUCTION

The requirement of plants for a certain duration of daylight (in order to flower) is now well known. In order to flower many plants require—in addition to a low temperature period—a short photoperiod of 10 hours or less, and others require one of 14 hours or more. The light stimulus is perceived by the leaves and the interval over which the response to day length occurs is that of photoperiodic induction. For short-day plants, the dark period is the critical one if it has been preceded by a light period. Long-day plants flower even in continuous light and then failure to flower is due to an excessively long dark period. Some plants require a succession of two different photoperiods. Thus *Cestrum nocturnum* flowers when subjected to long days followed by short days or continuous long or short days or short days before long days are ineffective.

The behaviour of short- and long-day plants is presently explained in terms of flowering hormones. Florigens, formed in leaves, would be destroyed by high levels of auxins. But when florigens are translocated to buds, auxins contribute to their fixation and to the subsequent differentiation of floral primordia. In floral induction, according to Liverman (1955), there is a lowering of the auxin level during the dark period in both long- and short-day plants. In short-day plants, the dark period causes the auxin content to drop to a level where florigen synthesis can occur. In long-day plans, the auxin level is too low for flowering under short-day conditions and long days are thus required for floral induction.

When occurring late in the life of a plant, photoperiodic stimuli sometimes cause abnormal responses. Thus short-day conditions may cause the formation of embryo sacs in the anthers and of pollen in the pistils.

Even plants indifferent to day length, such as tomato, fail to fruit when exposed to day lengths of 5 hours or less or of 24 hours or more. In the latter case, foliar injury often occurs.

Photoinduction occurs both in woody and herbaceous plants. *Poinsettia* and *Bougainvillaea* are short-day woody plants and *Hibiscus syriacus* is a long-day plant. Other species are day neutral. Although in herbaceous plants, photoperiods induce flowering, in woody plants they control dormancy primarily. Sometimes, as in apple flowering, climatic conditions are less important than physiological conditions and the latter are modified by climatic conditions in ways which are still obscure. In tropical and subtropical plants, dormancy or vegetative pause seems to depend upon chilling or dry spells.

A number of growth regulators interfere with floral initiation, and Salisbury (1957) has identified the steps in photoperiod induction of cocklebur that are blocked by some growth regulators. One mechanism determines the critical day length. Other stages of induction are synthesis of flowering hormone in the leaf and development of floral buds. Evidence for this reasoning is based on the ability of cobaltous ion to interfere with the mechanism controlling critical day length; the inhibition by 2, 4-dinitrophenol of florigen synthesis; the action of 3-indolacetic acid, naphthalene acetic acid, and 2, 4-D to destroy the flowering hormone in the leaf; and the ability of maleic hydrazide, Dalapon, 2, 4-D to inhibit floral bud development.

Quite apart from photoperiodic effects, intense light absorbed by dark surfaces such as fruits and flowers is converted into heat and the exposed tissues can attain temperatures 8 to 9°C above the temperature of the surrounding air. Blueberries in New Jersey attained temperatures of 40°C when the air was 31°C.

Intense light prevents the germination of many seeds, e.g., most Liliaceae, and induces secondary dormancy in some of them, e.g., *Nigella*, which become light hard. In such cases their germination is hindered, even when they are put again in darkness. Dark hardness may be caused by low light on light-requiring seeds, as, for example, many grasses.

In insufficient light, leaf production takes place at the expense of roots and reproduction is impaired. Delayed flowering, reduction in the initial number of flowers, shedding of blossoms, and slow development of fruits are usually induced by low light intensities, either at high or low temperatures.

When plants receive the full spectrum of daylight, intensity is not very important. About 40 foot-candles seem sufficient for mere survival of many plants. In Guthrie's experiments (1929), flowering was impaired only when illumination was reduced to 8% of full summer sunlight. When harmful effects occur, they usually result from an incomplete spectrum, which lacks the shorter green, blue, and violet wavelengths, i.e., shorter than 52,000 Å. Under glass that fails to transmit violet light, some plants differentiate few if any flowers.

The studies of Flint and associates were among the first to present modern information on the influence of different rays on the life of plants. According to them, the region 5200 to 7000 Å (red, orange, and yellow light) stimulates the germination of lettuce seeds, whereas the regions from 4200 to 5200 Å (green, blue, and violet) and 7000 to 8600 Å (mainly infrared) are inhibitive. The action spectrum for floral induction of short-day plants, such as *Xanthium pennsylvanicum*; is similar to that for germination of lettuce seed. Present data suggest that red and infrared rays generally participate in flower induction and elicit other morphological responses of higher plants; so that the action of light upon some seeds would be one aspect of a general phenomenon influencing living processes.

Red rays promote growth and flowering in barley, a long-day plant, whereas infrared rays inhibit growth and promote flowering in *Xanthium*, a short-day plant. In flower initiation, the most effective red rays are in the region of 6500 Å and 7350 Å is the most effective region of the infrared. Red rays may repeatedly reverse the photoreactions induced by infrared and vice versa. These reactions appear to be independent of temperature. Their reversibility, however, is reduced by a dark period between the two irradiation periods. The effect of darkness depends upon temperature.

Infrared rays may also decrease fruit set, especially at high temperatures, in some neutral day plants, such as tomato.

VI. INTERACTIONS OF NUTRITION AND CLIMATE ON REPRODUCTION

Although the effects of nutrition and climate on reproduction have been examined separately, their action on plants is sometimes inextricably interrelated and cumulative. Thus, long- and short-day plants do not respond to photoperiodic stimuli when temperatures are too low (1° to

4°C) or too high (30 to 38°C). In short-day plants, the night temperature is very important, and in long-short day plants, e.g., *Cestrum nocturnum*, the day temperature has a great effect upon short-day induction. During part of the photoperiodic cycle, cool temperatures are required by *Glycine max*, perhaps indicating an "endogenous rhythm" of 24 hours' duration.

The behavior of seeds is altered by combinations of temperature and light. Lettuce seeds, kept for 24 hours at 25°C in a dark germinator lose their sensitivity to light and do not respond to standard illumination. However, when stored in darkness at 5°C, they germinate normally. In a dry environment, seeds exposed to low temperatures do not after-ripen; but if allowed to imbibe water before exposure to red or infrared rays, the response of the seeds is different.

The nature of the flowering response is cucurbits is altered by proper adjustment of temperature and light. Thus, cucurbits can be induced to form staminate, monoclinous, and pistillate flowers under appropriately controlled conditions, just as they do in nature.

High temperature in association with low light intensity, induces poor fruit set in tomato as well as dormancy of fertilized ovaries. In some fruits, high temperatures, intense light, and deficient moisture, in combination, alter cell permeability with the result that cell sap fills the intercellular spaces, parenchymal cells become discolored, and a watery decay of tissues sets in, e.g., water core of apples. Blossom end rot of tomatoes is another example. It can be induced by intense daylight, by high night temperatures, by temporary water shortage in a critical stage of the development of the fruit, or by calcium deficiency in which the ratios of potassium to calcium and of soluble calcium to total salts in the soil solution are important.

"Shrunken grain" of wheat is another complex disease induced by moisture deficiency of the caryopses, which deficiency in turn is caused by hot dry winds, by pathogens such as rust or *Ophiobolus graminis*, or by malfunctioning of the vascular tissues, such as in lodging of wheat. According to Mulder (1951), bitter pit of apples is induced by inappropriate breeding and cultural methods, excess nitrogen, relative lack of phosphates due to magnesum deficiency, and adverse ratios of foliage to fruits.

VII. IMPAIRED GROWTH CORRELATION OF REPRODUCTIVE ORGANS

The fulfilling of each stage of reproduction requires that the different organs be in the right condition at the right time. Thus, for fertilization to occur, pollen must be able to germinate, stigmas must be receptive; water, inorganic nutrients, sugars, and growth factors must all be adequate; and temperature and light must be within certain limits. If these conditions are not fulfilled, fertilization may be impaired. In tomato, high nitrogen levels, or low light, or high temperatures and high light intensity may induce style exsertion before dehiscence of the anther sacs occurs and, as a result, fruits fail to be set. In maize, the environmental conditions may easily induce the opposite condition, premature pollen dispersal, though this is usually not very harmful.

VIII. PREMATURE ABSCISSION OF REPRODUCTIVE ORGANS

Histologically, when flowers and fruits drop prematurely, a dissolution of the middle lamellae and adjacent layers of the primary cell walls occurs in the abscission zone of the peduncle. Meristematic layers protect the stump of the peduncle, which remains on the plant, and the lumens of vessels become occluded by gums and tyloses.

Abscission is now thought to be controlled by the relative auxin concentration across the abscission zone. When the gradient of auxin across the ascission zone—from the proximal to the distal side—is steep, flowers and fruit are not shed; when the gradient is low, or disappears, or is reversed as a result of an auxin spray on leaves, abscission results. The "auxin gradient" hypothesis may not be generally valid, since in *Phaseolus vulgaris* and *Coleus blumei* the controlling factor in stimulation or inhibition of abscission is the total amount of auxin applied and not the auxin gradient.

In young flowers, auxin production is largely centered in the maturing stamens and ovaries. When flowers are mature, their auxin content becomes very low, but upon pollination the ovary produces a new flush of auxin. This gradually diminishes until new auxin is produced by the endosperm and later by the embryo.

In the apple, there are four flushes of blossom and fruit abscission. The first drop is given by aborted and unfertilized flowers. The second appears associated with an insufficient activation of auxin-forming systems in the fertilized ovaries. These two can easily fuse one into another. The third (June drop) and the fourth (preharvest drop) occur in periods of lessened auxin production in endosperm and embryo. Auxin sprays satisfactorily control the first flushes of flower and fruit drop, when some environmental condition has temporarily interfered with pollination and limited fruit set.

Environmental conditions, however, can interfere with auxin action. Thus, Marglobe and Rutgers tomatoes do not respond to sprays of *p*-chlorophenoxyacetic acid under high temperatures and light intensities of summer, whereas summer setting varieties, sprayed with the same compound, nearly doubled their fruit production. When there is a temporary lack of nutrients, as in the case of early tomatoes that may flower when there are still few leaves on the plant, auxin treatment can usefully delay fruit drop until nutrition becomes adequate, but auxins cannot indefinitely replace an adequate nutrient supply.

The competition for nutrients among reproductive organs may result in dropping of fruit shortly after fruit set. And, when the number of leaves is insufficient to supply the fertilized ovaries with the necessary food, the younger ovaries may become dormant because the fruit already set seems to have a priority upon the amount of nutrients and food available. In tomato and cotton, the dropping of young fruits is greater when older fruits have already been set on the plant. In cotton, for example, the dropping of fruits at the beginning of the fruit set period may be as low as 10%, whereas at the end of the flowering period, it may be over 90%. In this case, the abscission of younger fruits is stimulated by the auxin concentration of older fruits. Inasmuch as nutrients are translocated to and concentrated in regions high in auxin, young fruits receive fewer nutrients than older one do.

The activity of auxins is, therefore, interrelated with and affected by nutrition as well as by light, temperature, respiration, and translocation. Furthermore, calcium is necessary for the formation of the insoluble pectins in middle lamellae and zinc is required for auxin synthesis. Carbohydrates are specifically useful in the formation of cell wall material. Oxygen and carbon dioxide concentrations, insofar as they influence respiration at this stage, may influence abscission. The carbon: nitrogen ratio (a resultant of the processes of nutrition), photosynthesis, and respiration also provide information on the likelihood of abscission. Either a very high or a very low value of the carbon: nitrogen indicates that fruit drop is likely.

Auxin sprays are a convenient method of preventing flower abscission, application being made to flowers or to the soil. A number of synthetic growth hormones are useful for this purpose. Esters of *p*-chlorophenoxyacetic acid, and of β-naphthoxyacetic acid are useful for improving fruit set of tomato. Very low concentrations of 2, 4-D have proven effective on snap beans. Benzothiazole-2-oxyacetic acid, a weak auxin, is preferred for use on figs because it causes development of seed coats as well, and this has improved the saleability of the parthenocarpic fruit.

For the control of preharvest fruit drop, auxins are applied as sprays or aerosols and treatment is repeated. The treatment tends to improve coloring of apple fruits and does not increase the percentage of defective fruits. The use of auxins in this way has some unfortunate side effects. Thus, the respiratory rate of the fruit is increased so that the useful storage life is shortened. To counteract this effect, maleic hydrazide can be incorporated into the auxin spray. Auxins also enhance radial cracking in apples and induce puffiness and other undesirable effects in tomatoes. Unless the concentration of 2, 4-D on oranges is at the low level of 10 mg per liter is tends to induce abnormally large navels or seeded fruits in the otherwise seedless Navel orange.

Olives fail to respond to auxins sprays, applied to prevent preharvest drop. Melis (1949) has pointed out, however, that both preharvest and summer fruit drop of olive is often caused by infestations of *Prays oleellus*, the importance of which had been overlooked.

Other auxins have been used for fruit thinning. It is thought that they may act by inducing incompatibility between pollen tubes and stylar tissue or by inhibiting pollen germination or tube elongation.

IX. PHYSIOLOGICAL DISORDERS FAVORING THE ESTABLISHMENT OF PARASITES ON REPRODUCTIVE ORGANS

Physiological disturbances sometimes make the plant more susceptible to attack by parasites. Thus, intumescences of peas and other legumes, induced by high humidity, have been attributed to *Cladosporium pisi*, which only colonizes them. Grape infloresscences, injured by low temperatures, are usually invaded by *Botrytis cinerea*. Olive pistils are often infected by *Pseudomonas savastanoi* through wounds caused by late frosts. Sunburn or sunscald of tomato is readily infected by ubiquitous molds such as *Alternatia tenuis, Cladosporium herbarum*, etc.

The colonization of necrotic tissues by common molds is so inevitable that some physiological disorders have long been mistaken for infectious diseases. Blossom end rot of tomatoes is an example. In southern Italy, the damage caused by *Monilia laxa* to almond blossoms is so dependent upon cold and humid weather that farmers attribute the rot and shedding of blossoms directly to unfavorable whether. In some of these diseases, the action of the pathogen is primary in importance although secondary in order of time. Such diseases might, therefore, be classed as "consequential" or "concatenate" diseases ("Folgekrankheiten" or "Kettenwirkung verschiedenartiger Krankheitsprocesse").

X. INFECTIOUS DISEASES

Whether infectious diseases are systemic or local, they affect the general physiology of the host and, when the infection exceeds a certain intensity, reproductive activities of the host become affected. The same pathogen can behave differently in regard to host reproduction, depending on the mode and amount of invasion of host tissue. *Kunkelia nitens* prevents flowering of blackberry (*Rubus* spp.) when mycelium is generalized in host tissues from the rootstock to the growing point, but does not interfere with blossoming except in the invaded nodes when infection is local.

Some pathogens of flowers, fruits, and seeds limit their attack to accessory parts of reproductive organs and thus do not interfere with the preservation of the species. *Coryneum beijerinckii* behaves in this way because it injures the bulky or dry pericarp of stone fruits.

A. Seed-Borne Pathogens of the Vegetative Organs

Some parasites cause no injury to the seed but are simply carried on them and subsequently attack the plant developing from this seed. *Corynebacterium michiganense* plugs the micropyle and surrounds the embryo of tomato seeds, but may not interfere with the germination of the seed or with the life of the seedling. Thus, the disease is latent in the seed or seedling and becomes overt only in the vegetative organs of the adult plant. Similarly tomato plants, attacked by *Fusarium bulbigenum lycopersici*, often bear apparently normal fruits and viable seeds. In such cases, the reproductive organs act as carriers of a latent infection.

This situation is opposed to that of the parasite which generalizes in the green parts of the host plant and becomes pathogenic only in reproductive organs, e.g., many smuts.

B. Diseases of Vegetative Parts Specifically Affecting Reproduction

It is well known that parasites may stimulate distant responses, useful or injurious, on the part of their host plant. There are orchids (*Gastrodia elata*) that do not flower unless infected by *Armillaria mellea*, and *Rhizoctonia solani* induces the formation of aerial tubers on stems of potato plants infected in the roots and stolons.

Occasionally, responses unfavorable to reproduction are elicited by mycorrhizae, the importance of which in the normal physiology of many plants is well known. Mycorrhizae of the olive tree may intercept the major part of nitrogen available in a nitrogen deficient soil. Pistil abortion and blossom shedding ensue. Sometimes this nitrogen deficiency results in morphologically abnormal staminate flowers, as Petri (1942) and others have found.

In northern Italy, the nitrogen starvation resulting from mycorhizae of *Ruta graveolens* has prevented the development of the embryo in otherwise normal seeds so generally that extinction of the species seems inevitable. Plants lacking mycorhizae produce germinable seeds.

The indirect action of pathogens on reproduction is illustrated by *Gibberella fujikuroi*, which forms gibberellins in the plant. This fungus induces earliness in plants reaching adult size, but lowers yields. Generally, the gibberellins increase stem elongation and, in doing so, release a flowering response, especially evident in plants where the acceleration of vegetative growth eliminates mechanical barriers to flowering. The gibberellins may reduce germination time of seeds, but have no effect on total germination.

C. Pathogens Attacking Reproductive Organs Directly

1. *Seed and Fruit Contaminants.* Parasitic or saprophytic microorganisms may get into the flower and be borne on the surface of the seeds as external contaminants; or they may penetrate into its external layers, inducing "transitional infections", "pseudoflower infections", or "flower-seedling infections." These microorganisms interfere with the viability of the embryo only when favored by particular environmental conditions. Among the genera of fungi acting in this manner are *Alterna-*

ria, Cladosporium, Helminthosporium, Pullularia, and *Stemphylium.* "Heating" of moist grains is due to these fungi.

Most of the parasitic fungi that act in this way infect the seedling during preemergence or immediately thereafter. Some, e.g., *Helminthosporium gramineum,* are also harmful later in the life of the plant.

2. *Fructicolous Infections.* Some pathogens attack reproductive organs only. Thus the flower primordia of *Pennisetum* are attacked by *Tolyposporium penicillariae*, monoclinous flowers ready to blossom are attacked by *Claviceps purpurea*, and entire female flowers of *Ulmus* are attacked by *Taphrina* spp. Pistils are also attacked. Sometimes they are partially infected, as in "partial bunt" of wheat caused by *Tilletia indica.* In others they are entirely invaded, as in the case of mulberry ovaries attacked by *Ciboria caruncoloides*. The seeds of *Alnus* are attacked by *Helotium seminicola* and fruits may also be attacked as in the case of *Tilletia barclayana* of rice.

These parasites sometimes enter sexual organs through the accessory parts of the flower or fruit. In other cases they infect the pistils directly or are inoculated by insects into seeds or fruits, e.g., stimatomycoses.

Many pathogens overwinter in the affected organs of the host, some in the soil by their own perennating structures, as in the case of *Claviceps purpurea* and many smuts; still others survive in ways that are not yet known (stigmatomycoses). These pathogens are not externally seedborne, do not induce generalized infections of their hosts, and might be considered "fructicolous" *sensu lato*.

3. *Generalized Parasites with Sporulation Restricted to Reproductive Organs*. Other pathogens, perhaps the most interesting for students of diseases of reproduction, are more or less generalized through the green organs, where their parasitism is more or less indifferent to the host plant. They sporulate and, in doing so, induce overt disease in reproductive organs or in the accessory organs of flowers and fruits. Gäumann (1950) considers these as agents of "organotropic diseases."

In rare cases the infection begins from the embryo, which becomes infected at anthesis. *Ustilago nuda* and *U. tritici* provide examples of this condition. Gäumann (1950) has used the term "germinative transmission" to describe this situation and Fischer and Holton (1957) used the term "embryo infection." In other cases, infection may begin in the seed coat, e.g., *Botrytis anthophila*; in other instances, it arises from the pericarp and husks, e.g., *Ustilago avenae*, from the seedlings from the shoots, as in *Ustilago violacea*. The period of incubation of shoot infections may last more than one year in plurennials. Parasites causing seedling infections and those causing seed coat, pericarp, and husk infections have seed-brone or soil-borne spores or mycelia. These situations correspond to Gäumann's (1950) terms "pseudoflower infection", "flower-seedling infection", and "adherent transmission." Such modes of infection may coexist in the same species. Thus, *U. violacea* can penetrate into the plant through flowers, buds, and excised stems.

The chrysanthemum aspermy virus causes deformation of flowers but no symptoms or at most a slight mottling on leaves. The virus can be recovered from roots, stems, leaves, and flowers; sap from flowers is the most infective, that from roots the least. Leaf sap is said to contain an inhibitor, although little or none is present in the flowers. Therefore, this is a case of "forced organotropism."

The diseases discussed in Sections X, C, 2 and 3 have been termed "venereal diseases" by Gäumann (1950). This is a suggestive term. Venereal diseases, however ("venereal" from Venus, the goddess of love) are not such simply because they affect the reproductive organs, but because they usually arise from sexual intercourse with an infected mate. Thus cancer of the reproductive organs is not a venereal disease. In the diseases considered above, the pathogens infect the reproductive organs but this is independent of the fertilization processes which are often prevented by the diseases caused by the pathogens.

In some cases, they effect also a "parasitic castration" (*Ustilago violacea*) or induce a "parasitic unfruitfulness" in a wider sense. Commonly, they do not penetrate into the host through the stigma and the stylar tissue. Even *U. tritici*, which shows so high an electivity for embryos, pierces the wall of the young ovary laterally.

It seems interesting to note that an apomictic progeny of a male sterile mutunt of *Paspalum notatum* produced more than five times as many ergotized florets, infected by *Claviceps paspali*, as a similar progeny of a male fertile sister plant tested in comparison. The sphacelial stage of *Claviceps purpurea* developed even on florets of wheat deprived of stamens and ovaries, *Tilletia buchloeana* and *Sorosporium Everhartii* may transform into bunt or smut balls the rudimentary ovaries of staminate spikelets, which normally would never produce seeds. Therefore, these fungi are to a certain degree independent from the presence of mature sexual organs and from the nutritive environment including hormones and similar substances associated with the reproductive state.

A similarity to venereal diseases may be shown only by those diseases which are transmissible by way of the fertilization process. This is the case of bean mosaic virus (*Phaseolus* virus 1), which can be transmitted through the agency of the pollen of infected plants. This virus diseases, however, is also transimitted in other ways and, when transmitted through the pollen, does not particularly damage reproduction.

4. *Reproduction Affected by Nonelective Parasites Entering the Plant through Reproductive Parts.* Still other parasites (*Erwinia amylovora* and *Sclerotinia* spp., especially *S. laxa*) are capable of attacking tender vegetative organs, such as unfolding buds and young leaves or twigs through natural openings and through wounds, but they frequently penetrate the plant through floral organs such as petals, stamens, and pistils. These structures are either not cuticularized or little so, and offer conditions conducive to infection; high humidity and the presence of nectar and stigmatic secretions. Ubiquitous organisms may also infect stigmas and enter fruits where they usually remain ineffective, as, for example *Bacillus vulgatus* in *Cucurbita pepo*. High atmospheric humidity favors the infection process. Thus, according to Weaver (1950), at 90% relative humidity peach blossoms contract infection by *Sclerotinia fructicola* through all floral organs; at 70 to 80% relative humidity, blight occurs only from infected stigmas; while at 70% relative humidity, blight of blossoms occurs only from stamen infections.

In the stone fruits, flower buds open before foliar buds do, and when the season is most favorable to infection, floral infections are abundant. In other cases, as with *Erwinia amylovora*, flowers attract the insect vectors of the pathogens rather than the pathogens themselves.

Some pathogens may cause late infections on maturing fruits, entering through lenticles or by piercing the cuticle as, for example, *Sclerotinia* spp. In these fruits, which are changed into mummies, the perfect stage of the pathogen is differentiated. Thus, fruits infected late act as primary host organs and flowers and vegetative parts as secondary host organs.

D. Undesirability of Sharp Discrimination between Vegetative and Reproductive Organs as Substrates for Pathogens

Gäumann (1950) rightly points out that sexual organs are the best individualized organs of the plant body. This individuality refers not only to the morphology and to the tissue characters of the reproductive organs, but also to the microenvironment they offer, which is peculiar and favorable to most pathogens, elective or not. Furthermore, in annual plants and in cereals in particular, seeds are the organs which permit the non-pernennating structures of the parasites to survive best. Several examples have already been given to show that the suitability of reproductive organs for pathogens is not uniform through the flower and fruit.

Many pathogens affecting flowers, fruits, or seeds are able to sporulate on or to infect certain reproductive organs only, such as stamens and pistils. Among these pathogens are some of the least specialized which are provided with the grossest symbiotic adjustments. Electivity may be so strict that in some instances infection or sporulation can occur only on stamens, as, for example, *Ustilato violacea* and *Botrytis anthophila*, or on stigmas and filaments, e.g., *Peronospora stigmaticola*, or on peducles, e.g., *Mycosyrinx cissi*. Others are restricted to petals, e.g., *Peronospora corollae*, seed funicles and placentae, as is the case for *Ustilago duriaeana* and *Schroeteria* spp. Viennot-Bourgin (1953) distinguished two types of floral galls induced by the Ustilaginales. In the first type, chlamydospores arise only in the stamens, the external flower verticillia not showing evident anomalies. In the second type, the whole of the floral organ, except stamens, participate in chlamydospore production.

Most parasites which overwinter as resting mycelium in viable seeds or fruits do not invade or injure the embryo, and *Ustilago tritici*, which invades the embryo, is elective for the scutellum and the growing point region and does not penetrate into the radicle. As stated above, even parasites showing a less high degree of parasitism may not behave indiscriminately toward the different parts of the flowers and fruits. *Ustilaginoidea virens* does not affect stamens and pollen grains, and the filaments of the flowers of sour cherry are infected by *Sclerotinia laxa* (which displays poor electivity) only through wounds.

All the organs of the flower are affected by *Claviceps purpurea*, which parasitizes even the remnants of the glumes in heads invaded by *Ustilago tritici*, but *C. purpurea* tends to form sclerotia and consequently ergotamine and related compounds, only in the presence of the ovary. This is detached from the receptacle and remains at the apex of the developing sclerotium.

The foregoing examples suggest that many parasites are elective for single organs, and the activity of these pathogens is independent of the reproductive activity of the host. In some cases, however, infection is coordinated with the fertilization process. *Claviceps purpurea* commonly infects wheat before, and *Ustilago tritici* after the fertilization of the ovule.

That the activity of the pathogen is, in fact, independent of reproductive activity may be shown by experimentation or through field observation by demonstrating that green tissues, which are normally not invaded or injured by the elective parasites considered here, are at least as subject or attack by these fungi as the reproductive organs. Some conditions favor such fungi as *Tilletia caries, T. foetida*, and *Ustilago tritici* to sporulate on vegetative organs, although these fungi usually manifest themselves in the inflorescence of wheat. The greenhouse environment, the amputation of the inflorescence, and the infection of lemmas by *Trichothecium roseum* or *Fusairum culmorum*—all favor this type of action. Viennot Bourgin (1937), therefore, thinks that the parasitic action of *Ustilago nuda* f. sp. *tritici* makes it possible to approximate the Ustilaginales localized in vegetative organs to the naked smuts, usually considered specific for the inflorescences; and that a transition exists between the smuts sporulating on the ovaries, on the flower involucres, and on leaves (*Ustilago maydis*).

The best known fructicolous fungi show similar aspects and the conclusion of Cherewick (1953) about *Claviceps purpurea* is that ergot may develop not only on young ovaries but on any physiologically young tissue of the wheat and barley plant.

Sometimes the behaviour of the pathogen is more complex. It changes, although the host organs and environmental conditions are equivalent. Maybe in these cases, the stage of the disease reveals its importance. Why do some Ustilaginales, e.g., *Tilletia barclayana* and *T. indica*, soon sporulate in ovaries on which their wind-borne conidia alight, whereas others, e.g., *Ustilago tritici*, sporulate only in the flowers they enter from adjacent vegetative tissues that are generally invaded? What changes does the host-parasite complex go through during the development of in infected wheat plant?

Many parasite affecting reproduction enter the plant through its vegetative organs and can be controlled by protective fungicides or by the natural defenses of the host during the process of infection. We might, therefore, conclude that the behavior of the green parts of the plant is of even more interest than that of the reproductive parts on which the pathogens usually manifest themselves.

XI. THE INFLUENCE OF REPRODUCTION ON THE ACTIVITY OF PATHOGENS AND DISEASE EXPRESSION

The stimuli for the transition of the plant from the vegetative to the flowering condition excite responses throughout the plant, and influence the plant in its entirety. Therefore, even in diseases not especially related to reproduction and to the organs directly responsible for it, maturity and reproductive activity often upset the equilibrium between host and pathogen.

In some cases, the flowering plant overcomes its invaders, as in the case of root nodule bacteria on legumes. More often the reverse is true. The attack of mature plants by rusts, the epidemic phase of *Phytophthora infestans*, the leaf target spot stage of *Alternaria solani*, the *Oidia* as agents of "Alterskrankheiten", all afford examples of microorganisms more or less favored in this way. Genetic, seasonal, ecological, and nutritive factors, separately or in association, can also play a role in these changes of the physicochemical environment of host tissues. Recently, Grainger (1957) has proposed an index for the interpretation of the differences in susceptibility of new sprouts young plants, and mature potato plants to *Phytophthora infestans*. This index indicates susceptibility or resistance according to its high or low value and consists of the ratio of the weight of total carbohydrates in the whole plant, C_p, divided by the residual dry weight of the shoot, R_s. The ratio $C_p : R_s$ has been usefully adapted by Grainger for other diseases as well. The study of tissue factors, which are directly responsible for arresting the process of infection or for allowing it to evolve into over disease, is still in a most primitive state in animal and plant pathology.

It is sufficient to conclude here that young vegetation may resist infection or keep it latent and be a "carrier" of disease, which becomes overt when the plant is mature. In such cases, the behavior of the young plant can be opposed to that of the same plant when adult. In these cases, therefore, the determinants of the pathological process are not the infectious agents, but the physiological changes of the host plant at maturity.

From the very different point of view, reproduction is a useful tool in checking the spread of and the economic effects of infectious diseases. Vegetative propagation results in genetic uniformity. Fertilization, and especially cross-fertilization, by contrast, makes for genetic heterogeneity. Stevens (1948), analyzing disease damages in clonal and pollinated crops, found disease was more frequent on the former than on the latter and that losses were least in cross-fertilized plants.

XII. REPRODUCTIVE ORGANS AS INJURIOUS AGENTS

Occasionally, pollen grains are harmful to human beings and injure vegetation, too. Valleau (1949) demonstrated that the leaf bleaching of tobacco growing near corn is caused by corn pollen on the tobacco plants. The reasons for this response are unknown.

6

EFFECT ON PLANT SENSITIVITY

The term hypersensitivity is borrowed from human medical terminology. Literally, it means that an organism or group of organisms is sensitive to a pathogenic agent beyond the norm. As can be seen from the following amplifications, the definition describes only one aspect of the phenomenon here under discussion. Moreover, the hypersensitivity concept in plant pathology differs essentially from the same term in human medicine. Thus, no analogous conclusions can be drawn from examples in animal pathology.

What does the phytopathologist understand by the term hypersensitivity and its synonyms (suprasensitivity, hypersusceptibility, hyperergy, Überempfindlichkeit, and in connection with these concepts: incompatibility reaction, protoplasmic resistance, active resistance, etc.)? It is suitable to a preconceived terminology.

The Cambridge botanist, Ward, was the first to recognize clearly the significance of the hypersensitivity reaction as a defense mechanism of the plant against potential parasites. In his treatise (1902) "On the relations between host and parasite in the bromes and their brown rust" he shows that the pathogen penetrates with its hyphae into resistant as well as susceptible hosts. He observed no differences between behavior of the resistant and susceptible hosts until direct, physiologic contact is established. In the congenial host, "the parasite slowly taxes its host and even stimulates the cells for some greater activity." In the uncongenial host, the pathogen induces changes that were described by Ward as follows: "The tissues turned brown and died, the destructive action of the infecting tubes having killed the cells too rapidly." At the same time, the pathogen ceases to grow. Thereby, the infection is, limited to a local necrosis and the plant escapes the disease. The writer attributes the premature physiologic breakdown of cells to an increased sensitivity of the tissue of the hosts toward some metabolic products of the parasite.

Ward recognized that both extremes "highest sensitivity" and "highest tolerance" of the host cell are connected by intermediary steps and furthermore, that the mode of reaction is influenced by environmental factors. He tried in vain to answer the problem, not as yet solved, why a pathogen such as *Puccinia dispersa* discontinues its growth prematurely in the hypersensitivity host. He endeavored to find "some body" in the cell sap of the host plant which inhibits or promotes the growth of the fungus. These efforts failed. "Of course", he emphasizes, "the experiments teach very little." Finally, he suggested that "internal, i.e., intraprotoplasmic properties beyond the reach of the microscope and similar in their nature to those which bring about the essential differences between species and varieties themselves" must the cause for the gradually changing behavior of the host plant.

About 10 years later, research on the phenomenon described by Ward received further impetus from Stakmank. His discovery that Eriksson's "formae speciales" of the cereal rust fungus consist of many physiologic races, the host spectra of which are determined by the interaction mechanism described by Ward, gave direction to further progress in hypersensitivity research. The term, hypersensitivity, was introduced by Stakman into phytopathological terminology.

Since the hypersensitivity phenomenon was first studied on cereal rusts, the opinion prevailed until the 1920's that it is connected with plant diseases elicited by obligate biotrophic fungi. The causal interpretation of the mechanism of the process was also strongly influenced by this opinion. However, it was slowly recognized that facultative parasites, such as *Phytophthora infestans* or *Venturia inaequalis* can bring forth reactions in their hosts that have all the earmarks of the phenomenon described by Ward for the rust diseases. Virus research provided a further impetus, since it was found that viruses may multiply within a short time even in the uncongenial host, but they induce necrosis in the infected tissues, causing a localization of the infection.

Finally, and most recently, the hypersensitivity or hypersusceptibility concept has been applied to interactions of some sucking insects and their hosts, that lead to a premature necrosis of the sucked part of tissue and that prevent the establishment of a normal parasitic relationship.

In looking back on the history of hypersensitivity research three ways can easily be recognized by which attempts were made to gain insight into the nature of this phenomenon. There are numerous works at hand in which scientists tried to find by histological or tissue-physiological methods how the course of interaction differs in "hypersensitivity" from the "normosensitive" hosts (structural analysis). The second way consisted in examining how far the interaction is determined by the genetic constitution of the host or pathogen and by environmental conditions (conditional analysis). And finally, attempts were made to find through physiological experiments and chemical analysis the factor that prevents the pathogen from developing in the tissue of the hypersensitive host after initial normal development (causal analysis).

Such separation can of course be maintained in theory. Actually, the approaches constantly overlapped, especially when histological findings were used to interpret the phenomenon.

In order to avoid misunderstandings, some definitions are expressed as follows:

1. Hypersensitivity reaction (HR) encompasses all morphological and histological changes that, when produced by an infectious agent, elicit the premature dying off (necrosis) of the infected tissue as well as inactivation and localization of the infectious agent.

2. A plant which does not react in a hypersensitive fashion and, therefore, is susceptible to the pathogen, is called "hypersensitive." Depending on the intensity of necrotic changes with which the cell of the host responds to the infection, distinct degrees of sensitivity are distinguishable.

3. Interactions that lead to a rapid death of the cell of the host and simultaneously to inactivation of the pathogen are called "parabiotic" according to Gäumann (1948). On the other hand, in case of a parasitic relationship, in which the cells of the host and the pathogen remain alive for a longer period of time (mutual tolerance) the interaction is called "eusymbiotic."

4. The expressions "resistance" and "susceptibility" are used exclusively in a clinical sense. Thus, "resistance" only means that a certain plant—notwithstanding the presence of a particular disease producing agent and conditions favorable for infection—remains free or nearly free of the disease. The term "tolerance" is applied to hosts (e.g., latent virus carriers) that do not become ill despite clinical infections.

5. The terms "virulence" and "avirulence" are used, respectively, with "susceptibility" and "resistance." They always refer only to the underlying host/pathogen combination (H/P combination).

It is impossible to define exactly the term "hypersensitivity" itself, for we know by experience only that in the resistant host the interaction occurs with greater force than in the susceptible one. The conclusion that the resistant host must be "more sensitive" than the susceptible one to some material influences of the causative agent is no more than a supposition still in need of experimental study.

I. HYPERSENSITIVITY REACTION AS MORPHOLOGICAL PHENOMENON

Hypersensitivity interactions produced by fungal pathogens will be discussed first, because we are best informed about them, and because the results achieved through them have decisively influenced the course of research.

A. Hypersensitivity Reactions Induced by Fungi

1. *Obligate Biotrophic Fungi**. Apart from Archimycetes, that develop entirely within the cell of the host, intercellular or extramatricular hyphal growth is characteristic for obligate biotrophic fungi. The cytoplasm of the cell of the host comes almost exclusively into direct contact with the hausteria of the pathogen. Since it is the eusymbiotic combination where the pathogen and host apparently live together "peacefully", a mutual tolerance between the border areas of both partners should be postulated.

Puccinia type: According to Stakman and Levine (1922), at least 6 interaction type can be differentiated in rust diseases of cereals for which thorough information is available:

i = no macroscopically recognizable symptoms

*The following definitions apply to this chapter: Biotrophic fungi are defined as those able to grow exclusively in living tissue (e.g., rust fungi). Fungi thriving in living tissue without killing it but able to grow on dead substrata are referred to as facultative biotrophs (e.g., smut fungi). The term necrotrophic is reserved for pathogens that spread in host tissues, killing it if the infection is not halted by the hypersensitive reaction (e.g., *Colletotrichum spp.*).

0 = chlorotic or necrotic speaks surrounded by a chlorotic rim area, no fructification of the causative agent

1, 2, and 3 = chlorotic and more or less necrotic specks with gradually diverse fructification of the causative agent

4 = no necrotic changes in the area of infected tissue, chlorosis only weakly indicated, optimal fructification of the causative agent.

A further interaction type, "x", was added to these six. It is characterized by the fact that three or more reaction types are realized on the same leaf (mesothetic types).

Extensive literature is available about changes that occur during interaction. The following results are considered essential in this connection.: The pathogen can penetrate into the plant, independently from interaction type. In infection type "i", interaction ends in a necrosis with only microscopically recognizable cell complexes. Scarcely developed mycelia are hidden in these; the development of a gum-like substance in the necrotically changed tissue is characteristic for this interaction type and for type "0." Mycelia advancing into the mesophyll, and penetrating with their haustoria into cells, are found in reaction type "0" that has been very extensively studied by Allen. After a relatively short period of incubation, "degenerative" changes occur in the invaded cell. The cytoplasm is condenses around the nearest haustorium. The physiological breakdown is indicated by progressive disintegration of plastids and cell nucleus, brown discoloration of cell contents, and swelling of cell membranes. Sharply defined, scorched-looking flecks are the end result. At the same time changes occur in the hyphae that indicate "exhaustion" of the pathogen: the hyphae show an increased affinity for nuclear stains, cell nuclei disintegrate into a homogeneous mass, density of the cytoplasm simultaneously increases in the haustoria-mother cells.

In eusymbiotically interacting H/P combinations (reaction type 4) the interaction fares differently. Degenerative changes in pathogen as well as host occur relatively late, about the time when the pathogen achieves fructification. Interaction types 1 and 3 occupy an intermediate position. In individual cells, necrosis—as far as its strength is concerned—is characteristically subject to a much greater variability in a narrower space than in type 0 or 4.

In principle, a similar course of HR can be observed in other rust diseases, such as flax rust, corn rust, or bean rust. The same is true for downy mildews, e.g., *Plasmopora viticola* and *Peronospora manshurica.*

Erysiphe type: As far as the available literature shows, mode of interaction and frequency of infection vary here too, one independent of the other. As Salmon already recognized the fate of the pathogen is determined only after the infection peg has penetrated into the cell of the epidermis. According to Neger (1923), who studied the host spectrum of *Erysiphe cichoracearum* and *E. polygoni* isolates, a premature collapse of the invaded epidermal cells occurs. The various H/P combinations seem to react differently. According to White and Baker (1954) and Hirata (1958) it is not the invaded cells but the adjoining mesophyll cells that react with necrosis in cereal mildew. Furthermore, White and Baker established a positive correlation between speed of reaction and degree of resistance. The sooner the collapse of mesophyll cells occurs, the weaker the development of extramatricular mycelium.

Synchytrium type: The behavior of potato toward *Synchytrium endobioticum* may serve as an example. Again, there is no difference between normo- and hyper-sensitive types in frequency of infection. It is found only in the degree with which the cell of the host "tolerates" the pathogen. Only meristematic tissue is prone to infection. It is characteristic that the tissue adjoining the infected cells in the eusymbiotic H/P combinations, responds to infection with development of tumors. In case of highest sensitivity the infected cells die after a few hours; so does the pathogen. In the intermediary reaction types, the cell of the host and the pathogen remain alive for some time. But, before there is tumor development in the tissue surrounding the infected cell, necrotic changes set in and lead to a cutting off and expulsion of the infection focus. In a borderline case, a great deal of cell multiplication continues in the neighboring tissue, but here too, abortion of the infection focus occurs, before the sori of the parasite have ripened. The normosensitive or susceptible host types is the opposite of these more or less hypersensitive host types. Here the cell of the host as well as the tissue surrounding the infection focus remains alive. The tissue develops into extensive tumors and the pathogen can multiply.

2. *Facultative-Biotrophic Fungi. Phytophthora type:* This type shows, histologically speaking, numerous similarities with the *Puccinia* type. The pertinent examinations have been almost entirely carried out with potato (interspecific hybrids) and *Phytophthora infestans. Phytophthora infestans* develops haustoria, as numerous obligate biotrophic parasites do that remain alive a relatively long time in eusymbiotic H/P combinations. Again there is no connection between the reaction type and infec-

tion frequency, either. The leaves react to the infection with rapid necrosis. The collapse of the cell of the host precedes the dying off of the pathogen, but its span of life in the necrotic tissue varies in different H/P combinations.

In the tubers, the interaction is considerably slower than in the leaves. If the infected cells show signs of an initial physiological breakdown after a 24-hour incubation period (appearance of melanins, regression of turgor, etc.), the pathogen can penetrate only as deeply as a few cell layers into the parenchyma. Under such conditions, no fructification of the fungus takes place. Since both extremes "highly susceptible" and "highly resistant" are connected by intermediary steps, no alternative separation into "susceptible" and "resistant" is possible in tubers. In the highly susceptible tubers, the tissue attacked by the pathogen remains alive for weeks and the haustoria of the pathogen exhibit especially strong growth, and the sheaths that surround them reach maximum strength.

Ustilago tritici type: In eusymbiotic combinations, systemic infection of the host plant is characteristic for this type. In the first type infection becomes evident only when the host flowers. Such a course of development presumes extensive mutual tolerance between both interaction partners. The other extreme is realized when, as in smut of oats caused by *Ustilago kolleri*, development of the pathogen stops in the first stages of infection. The invaded cells of the host collapse, and the pathogen stops growing simultaneously. As Oort (1947) and later Király and Lelley (1956) have established, the intermediate degrees of sensitivity are characterized by strong reduction of tillering and longitudinal growth. The intermediary types show greatly changing behavior in clinical results. In some plants the pathogen may grow into the flower region, in others it fails to reach the inflorescence, and the plant remains "free of disease", clinically speaking.

3. *Necrotrophic Fungi*. The high degree of mutual tolerance between host and pathogen in the presence of susceptibility is characteristic of the above described interaction types. In the necrotrophic pathogens, on the other hand, there is not a eusymbiotic relationship. If the HR mechanism fails, the pathogen produces a more or less rapidly developing decay of the attacked tissue.

The combination *Pyrus malus* and *Venturia inaequalis* should still be considered as a transition type. According to Nussbaum and Keitt (1938) and Shay and Williams (1956) the highest degree of susceptibility is characterized by a normal appearance of epidermal cells 9 to 10 days after infection by the subcuticular route. The pathogen spreads freely below the cuticle of the host and fructifies before the cells have finally collapsed. On the other hand, the impending physiological breakdown can be foreseen soon after infection in the cells of resistant forms of apples. Infection foci of neighboring cells also show symptoms of an initial necrosis. In the case of highest sensitivity, infection remains limited to a small group of cells, and no more fructification of the fungus takes place. Nussbaum and Keitt describe a third mode of reaction in which the development of a weak mycelium occurs. However, the cells that have come into direct contact with hyphae show no necrotic changes.

In the following H/P combinations, no eusymbiotic relationship, even a temporary one, seems to occur: *Phaseolus vulgaris* and *Colletotrichum lindemuthianum, Cucumis sativus* and *Corynespora cucumerinum, Limum usitatissimum* and *Colletotrichum linicolum, Hordeum sativum* and *Septoria passerini, Hordeum sativum* and *Helminthosporium gramineum, Zea mays* and *Helminthosporium maydis*. All these combinations have the factor in common that the pathogen can penetrate into a susceptible as well as a resistant host, and that in case there is resistance, a necrosis progressing rapidly in the infected host cells halts the establishment of a pathogenic relationship prematurely. On the other hand, in the susceptible host, while spreading from the site of infection the pathogen destroys the tissue of the host, often in advance with the dissolution of middle lamellae. Generally, the extremes are connected by intermediary reaction steps. The histological pictures often give the impression that the reaction depends primarily on the speed of the "disintegrating"—i.e., pathogenic—action of the fungus and the speed of the counteraction by HR of the host whether the spreading of the pathogen can be arrested in time.

More often than during attack by biotrophic pathogens, the host develops more or less marked "demarcation tissues" that shut off the infection from the healthy tissues. An instructive example is the shothole syndrome, described in detail by Neaf-Roth (1948), that is produced in *Prunus* species by infection with *Clasterosporium carpophilum*, and other leaf inhabiting pathogens. Basically, this appears to be a hypersensitivity reaction of intermediary degree, in which the spreading of the pathogen is halted relatively late. Then, approximately 20 cell rows away from the necrotic tissue complexes, the cells begin to divide and bring about the development of a concentric periderm for the separation of the necrotic tissue.

The hypersensitivity reactions produced by leaf pathogens were the first ones to attract the attention of phytopathologists. The hypersensitivity reactions occurring in other organs have only recently become the object of intensive studies. Thus, Flentje (1957) proved for *Pellicularia filamentosa* (= *Rhizoctonia solani*) and Hooker (1956) for *Pythium* types, that the reaction of the invaded cells of resistant hosts is analogous to changes, which have all the earmarks of an HR, histologically speaking.

The examples cited up to now are hypersensitivity reactions that are produced by infection of the host with "adequate" pathogens. The question is, what happens when the spore of a pathogenic fungus lands on the surface of a plant that does not belong to the natural host spectrum of the pathogen? This question was raised as early as 1905 by Salmon. He found that *Erysiphe graminis* also penetrates into "wrong host plants", but stops growing there after a short initial development. Similar results were obtained by Corner (1935) for other mildew fungi (*Podosphaera leucotricha* or *P. pannosa*). Hori (1935) and Müller (1950) studied the behavior of the causal organism of late blight of potato and tomato, *Phytophthora infestans*, toward plants that are not known as natural hosts of this pathogen (among them Composites Papilionaceae, and Liliaceae). If the pathogen succeeds in penetrating the tissue of the host, the cells respond with rapid necrosis, and the pathogen simultaneously stops growing. Sproston (1957) obtained similar results with *Monilinia (Sclerotinia) fructicola, Alternaria tenuis* and *Botrytis alli* after transfer onto *Impatiens balsaminea.*

B. Hypersensitivity Reactions Induced by Viruses

If they do not exceed a certain size, parts of tissue that become necrotic after virus infections are generally called "local lesions." Since in many cases no virus can be isolated at some distance from the local lesion, this phenomenon also must be based on a mechanism that localizes the pathogen. The necrotic tissue, however, contains considerable quantities of active virus for some time. According to many authors virus synthesis stops when physiological breakdown begins. According to Rappaport and Wildman (1957), the spreading rate of local lesions is constant in certain H/P strain combinations of *Nicotinia glutinosa* and tobacco mosaic virus, so that there is no absolute inhibition of the virus in this instance. Furthermore, a close positive correlation was found between the size of local lesions and quantity of virus that can be separated from the necrotic tissue. This result should point to the lack of an inactivation mechanism. But by using a slowly growing strain, regression of the relative virus quantity was obtained, which indicates that considerable quantities of virus are destroyed or inactivated in the necrotic tissue. Such an inactivation mechanism was postulated by Zech (1952). This author found that the center of older local lesions contains no more active virus, while large quantities of active virus could still be found in the border areas. Thus, in certain H/P combinations, HR could elicit not only localization but partial inactivation of the virus.

As far as the histological changes occurring during interaction are concerned, the following observations should suffice: The first changes recognizable with the naked eye occur in the infected tissue relatively early—in *Nicotiana glutinosa* and tobacco mosaic virus after 30 hours. According to Bald the cell nucleus is affected first, then the chloroplasts, disintegrate and protoplasmic streaming ceases. Finally, the cell collapses. In cases studied by Ball, a new layer of cells is attacked by this degenerative process, every 4 to 5 hours. The span of time, between infection and dying off of an individual cell, is 6 to 20 hours.

The symptomatic picture of HR occasionally varies greatly, depending on H/P combinations, age of host plant, and external circumstances (see page 490). Considerable differences can occur even in the same organ. In *Nicotiana tabacum* and potato X virus (strain Bf) Köhler (1951) described three local lesion types occurring in the same leaf, that differ in the varied arrangement of the necrotic and still living chlorophyll-carrying tissue rings. These differences might be based on periodic fluctuations of cellular sensitivity, according to Köhler.

As in fungal infections, HR can be chiefly observed when the host spectrum of an individual virus is examined. Schmelzer (1957) studied the reaction of 589 species of plants toward the tobacco rattle virus. Of these, 158 species, almost 25% reacted with local lesions. Of these, many species were not closely related to the *Nicotiana* genus. No dissemination of the virus into the surrounding tissue was observed in all these cases.

C. Hypersensitivity Reactions Induced by Bacteria

But little information is available to indicate that HR is also involved in bacterial diseases. Such a possibility should not be discarded, however. First, the etiological and the symptomatological momentum has taken precedence in the research of bacterial diseases up to now. Second, in bacterial diseases, if compared to mycoses, the interaction between host and pathogen is much more difficult to observe *in vivo*. However, the examples that follow should indicate that in bacterial diseases too, defense mechanisms occur that can be compared to HR induced by fungal organisms. When inoculated onto inadequate host plants, phytopathogenic bacterial may multiply for the first time, but stop growth after a certain period of incubation. Since no morphological barriers are observed, resistance probably results from physiological incompatibility of host and pathogen. The fact that many plants respond to bacterial infection with local lesions, indicates a mechanism which has much in common with HR to fungal infection. This idea is supported by the quantitatively different behavior of cotton species toward *Xanthomonas malvacearum*. The leaf lesions are much smaller in the resistant types than in the susceptible ones. Furthermore, *Pseudomonas tabaci*, elicits leaf blight in tobacco, and the closely related—perhaps even identical—*Ps. angulata* elicits only "angular leaf spot." Finally, the results by Fuchs et. al. (1957) should be mentioned, according to which *Pseudomonas mores-prunorum (p = Ps. syringae)* produces "die-back" in the branches of *Prunus* species, and only local necrosis in the leaves. The necrotic tissues are eliminated in the same way as in the shot hole disease of apricots and other stone fruits, caused by *Clasterosporium carpophilum*.

D. Reactions Resembling Hypersensitivity

Painter (1951) discusses the reason why a plant is unable to serve as a host to a certain insect. Among other things, he mentions hypersensitivity or hypersusceptibility. After giving a number of examples favorable to this concept, he concludes that hypersusceptibility and the resultant apparent resistance is not offered as an explanation for any insect-plant relationship but may be involved, especially in regard to insects with sucking mouth parts.

Studies by Börner and his colleagues on the reaction of resistant *Vitis* spp. and their hybrids with cultured grape varieties susceptible to *Phylloxera vastatrix* can offer the best insight into the problem. The leaf tissue of highly resistant hosts reacts to the attack by the insect with locally limited necrosis; simultaneously, the formation of tumor-like tissue typical of the susceptible host does not develop around the site of the bite. Since the development of a tumor appears to be the prerequisite for normal development of the insect, the plant remains free of damage. Five different overlapping reaction steps are differentiated. There is a positive correlation between the reaction of leaves and of roots, but the latter react to the attack by the insect less violently than the leaves. The sites of the bite are shut off by an inner wound periderm from the remaining tissue. The necrotic tissue parts are later "cast off" during the heavy growth.

A reaction mechanism similar to the one just described is suspected to be the cause of resistance to nematodes. The dying off of animals in the resistant hosts is ascribed to a premature disintegration of the tissue around the site of penetration, that cuts the animals off from their food supply.

II. HYPERSENSITIVITY AS A GENETIC PROBLEM

Biffen (1907) was the first to show that the hereditary resistance of rusts in cereals follows the Mendelian laws. A second, even greater impetus came from Stakman's work. It was soon found that HR toward individual rust races is controlled by individual genes. From the wealth of available literature, the following important points should be stressed within the framework of this essay, by which *Triticum vulgare* and *Puccinia graminis*, and *Linum usitatissimum* and *Melampsora lini* can serve as models.

The hypersensitive behavior toward an individual physiological race of the causative agent is usually characterized by a monogenous heredity in which "hypersensitivity" is mostly predominant over "normosensitivity." As a rule, an individual gene controls hypersensitivity toward several races of the causative agent, but various genes can determine hypersensitivity toward the same race. Analogous conditions are found in the pathogen. When this one interacts, as in rust fungi, during the dikaryon phase, "virulence" is chiefly recessive to "avirulence." Here too, a whole series of genes is involved. In the flax rust, at least 25 genes each for the pathogen as well as for the host, are controlling HR. The actions conditioned by individual genes—i.e., reaction potentials—interfere in pairs with each other; thus, the HR can only function when R_1 (resistance of the host) meets Av_1 (avirulence of the pathogen) in the cell of the host, or R_2 with Av_2, etc. Flor (1955), who has made the most detailed studies of this problem, interprets the situation as "conditioned by specific pairs of genes, one in the host and the other in the pathogen" (gene-for-gene relationship).

If one assumes that dominance corresponds to the building up of a substance and recessiveness to the opposite, then HR is triggered off by the interaction of two specific substances that are adjusted one to the other, one of which originates from the cell of the host, the other of the pathogen. Catcheside (1949) therefore compared the interaction of these two hypothetical substances with the antigen-antibody reaction, as we know it in animal pathology.

Many findings obtained from H/P combinations analyzed in less detail fit easily into this general concept of the genetic basis for HR. Hypersensitivity usually proves to be the dominant trait even in combinations where the pathogen is not an obligate biotrophic organism. It is mostly controlled by a limited number of genes. This is also true for the heredity course of resistance to viruses that is characterized by local lesion reaction. However, dominance of hypersensitivity is not always manifested. According to Holmes (1934) and Weber (1951) and others, in interspecific *Nicotiana* hybrids, for instance, the ability to respond with local lesions to infection by various strains of tobacco mosaic virus is controlled by hereditary factors with an incomplete dominance. Such a change of dominance was observed mainly in breeding material that was obtained from crossing susceptible with moderately resistant types of barley. Perhaps the simple gene dose is not sufficient to respond to the attack by the pathogen with a HR when the plants are kept under 15°–25°C temperature. Müller's (1930) hypothesis was based on a similar conception according to which resistance of potato to *Phytophthora* races A and C is based on the cumulative effect of 4 alleles of the R_1 gene and that a certain gene dose is necessary to set an effective HR into motion.

In a whole series of cases, however, the polygenous inheritance of hypersensitivity toward individual races of the pathogen has been obtained; e.g., for interspecific *Vitis* hybrids and *Phyulloxera vastatrix*.

The methods of mutation research have been used to clarify the genetic base of HR, in addition to the genetic analysis of hybrids that were obtained by intra- and interspecific crossings. Freisleben and Lein (1942) obtained mutants of barley, resistant to *Erysiphe graminis*, by X-ray treatment. Bandlow (1951) found that resistance thus obtained is based on hypersensitivity. Analogous results were obtained by Frey and Browing (1955) who produced mutants resistant to stem rust (reaction type 0 to 2). Here again, hypersensitivity proved to be dominant and conditioned by a single gene. On the other hand, Keitt and Boone (1954) succeeded in changing the pathogenic potential of *Venturia inaequalis* populations by treatment with chemical compounds that produce mutations.

III. HYPERSENSITIVITY AS A PHYSIOLOGICAL PROBLEM

Physiological research has contributed doubly to the clarification of the mechanism on which HR is based: first, in clarifying the physiological changes that occur in the parabiotically reacting tissues, and second, by finding out how far external factors influence the interaction between host and pathogen.

A. Comparative Physiology of Hyper- and Normosensitive Tissues

Most studies of this problem are concerned with H/P combinations in which an obligate biotrophic organism (including viruses) acts as a pathogen. Indeed, those studies have simply contributed the essentials needed for understanding "obligate parasitism." But, the one-sidedness in the choice of test objects also elicited a one-sidedness in the causal-analytical interpretation of HR. Namely, the physiological changes occurring in the hypersensitive tissues of the host and their extensive and early corresponding inhibition of the pathogen is mainly interpreted as a simple nutritional relationship. Actually, the relationships are not as simple as that. In any case, an interpretation based on this has to fail when HR is produced by an organism which can be grown on quite a variety of artificial nutrient media.

Brown (1955) characterizes the situation as follows: "To say that *Puccinia graminis* cannot progress through dead tissue is a truism if one accepts the current definition of an obligate parasite, but it is by no means excluded that the agent which inhibits the growth of *P. graminis* on an unsuitable host is comparable to that which acts similarly on *Phytophthora infestans* in the cells of a resistant potato variety. At all events the various responses of potato tissue to attack by the blight fungus are much more open to analysis than are the responses of wheat to the rust fungus and so, as it is good practice in ascending a staircase to begin on the bottom step and from there to work upwards, the same procedure could well be applied to the study of parasitism."

1. *Time Relationships*. One of the essential characteristics of HR is the relative rapidity with which the cell of the host undergoes necrotic

changes in response to infection with the avirulent pathogen. The span of time between infection and final breakdown of the cells of the host varies, depending on the underlying H/P combination and external circumstances; in broad terms, however, as far as more detailed studies show, it is always shorter in parabiotically interacting H/P pairs than in the eusymbiotic ones. In potato (Ackersegen type) and *Synchytrium endobioticum*, for instance, collapse of the cell is observed after a few hours; the interaction is finished after 16 to 18 hours in the combination *Phaseolus vulgaris* and *Phytophthora infestans*, to be discussed in more detail in Section III. In rust infections, on the other hand, days pass before there is a complete breakdown of cells of the host in parabiotically interacting H/P pairs. However, it takes place much more rapidly than in eusymbiotic H/P combinations.

The most detailed quantitative studies of rate of interaction and behavior of resistance are available for potato and *Phytophthora infestans*. According to Meyer (1940) and Müller and Börger (1940), at least five phases can be distinguished in the morphological and physiological changes that can be observed in the parenchymal cells of tubers. The parabiotic combinations go through these five phases much faster than the eusymbiotic ones, the extreme being 36 hours (19°C). In eusymbiotic combinations, on the other hand, the infected cells remain alive up to 3 weeks and the whole tuber is destroyed. Between these extremes are types of medium interaction rate and resistance. There is a clear relationship between both characters: the earlier the first signs of necrosis are observed, the earlier the pathogen stops growth. In extreme cases the area of tissue that has been penetrated by the pathogen is limited to a few cells.

Ferris (1955) did similar studies on the HR in leaves. The differences between the resistant and susceptible plants, as far as the rate of reaction is concerned, were not so striking. But she found that "a necrotic response was evident in the resistant plants at most a few hours before such a response developed in susceptible plants." The first differences in the mode of reaction between sensitive and normosensitive tissues become evident almost immediately after the pathogen has penetrated into the tissue of the host. Thus, increased plasma current and Brownian molecular movement start in the hypersensitive cells 10 to 60 minutes after infection, in the normosensitive cells only after 120 to 180 minutes. Analogous, but not as great, differences were observed in the speed of necrotic breakdown.

The concept of reaction rate as a deciding factor of HR only makes sense when we assume that during interaction the living conditions for the pathogen "become worse." Unfortunately, we are informed only incompletely about the chemical-physiological side of necrosis. But the substantial changes occurring during HR have to differ from those that occur for instance during necrosis induced by wilting toxins. Otherwise one could not understand why common saprophytes, such as *Alternaria tenuis* in *Impatiens balsamina*, discontinue their growth shortly after having penetrated into the tissue of the host.

2. *Changes in Chemical Constituents.* An integrating symptom of HR is the rapid loss of turgor in the infected cells. The water released by them is mostly absorbed by neighboring living cells. Consequently, the latter swell and fill the space left by the dead ones. Evidently, substances are also transferred which produce characteristic reactions in the neighboring tissue, such as wound periderm buildup, swelling and migration of cell nuclei toward the cell wall bordering upon the necrotic cell, or vacuolization of cytoplasm. Before the final physiological breakdown of the cell of the host, the hydrogen ion concentration increases in the cell. It is characteristic of some parabiotic H/P combinations that the amylolytic enzyme system is put out of commission; e.g., in potato *tuber* and *Phytophthora infestans.*

Tomiyama et. al. (1956) studied the water, carbohydrate and protein content of eusymbiotic and parabiotic H/P combinations in potato tubers infected by *Phytophthora*. They found in the latter ones—contrary to the eusymbiotic combinations—during the early stages of infection an increase in the water-soluble protein portion, in starch, and in content of phenolic compounds. The situation changed later on. The writer concluded that the infection results in the synthetic reaction in the tissue of resistant varieties. This is a nonspecific reaction of the host tissue to the damage done by the pathogen.

According to Noll (1950), a great accumulation of silicic acid (magnesium silicate?) occurs at the site of infection in parabiotic combinations (reaction type 0) between wheat and *Puccinia glumarum.*

The reaction of infected cells to stains is different in parabiotically and eusymbiotically reacting tissues. Irreversible affinity of cell wall and cytoplasm for basic vital stains (e.g., rhodamine B or neutral red, 2 p.p.m.) occurs much earlier in parabiotically than in eusymbiotically reacting tissue complexes. The individual cell shows increased affinity for the stain first where it is in direct contact with intercellular hyphae or haustoria of the pathogen. Thus, the cell does not react as a whole to the infection.

Unfortunately, there is not much information about the chemical nature of brown pigments, the most marked symptom of HR next to cell collapse. Undoubtedly the pigments contain phenol groups. According to Meyer (1940) they are deposited as pholobaphenes mainly in the intermicellar spaces of the cell walls. Since the cell membranes react positively in advancing necrosis to phlorogucin-hydrochloric acid, some authors think there is also lignification.

Buildup of tannin-like substance often precedes the appearance of brown pigments. Since these have antibiotic qualities, some authors consider them "antibodies." Dufrenoy (1936) tried to clarify the immunological significance of this group of substances. Humphrey and Dufrenoy (1944) postulated that in *Avena sativa* and *Puccinia coronata* there is a causal relationship between coacervation of phenolic compounds at the site of infection and the "decomposition" of the respiratory system. When this is so great "as to prove rapidly lethal, the rust fungus no longer behaves as a parasite, but as a pathogen inducing necrotic spots characteristic of 'hypersusceptibility."

3. *Metabolic-Physiological Changes. a. Respiration:* As is generally known, an increase in the intensity of respiration occurs in the parasitized tissues. Attempts have been made to gain insight in two ways: First, by comparative studies in eusymbiotic and parabiotic H/P combinations of respiratory intensity or of the activity of respiratory enzymes separated from host tissue and, second, by studying whether through manipulation of external circumstances or treatment of tissue with respiratory inhibitors the reaction potential can be changed according to a proposed working hypothesis.

According to Millerd and Scott (1956), in parabiotic combinations of *Hordeum sativum* and *Erysiphe graminis* respiration is clearly increased at the beginning of interaction, but the respiratory intensity falls sharply again, and the growth of the pathogen stops simultaneously. In eusymbiotic combinations, on the other hand, the respiratory intensity was clearly higher than in the control plants, even 168 hours after infection. Hosts with intermediary resistance behavior showed an intermediary character in the respiration behavior. Samborski and Shaw (1956) and Shaw and Samborski (1957) found a similar relationship in wheats and *Puccinia graminis.* Japanese workers (e.g., Tomiyama et. al., 1956) have stressed that respiratory increase is an essential factor in HR.

Millerd and Scott (1955) reported that from extracts of leaves attacked by mildew they had isolated a heat stable dialyzable factor, which effects an increase in respiratory activity of healthy wheat leaves. Later, a phenolic compound was isolated from the raw extracts, which produces increased sensitivity toward the mildew fungus in susceptible plants.

Sempio (1950) tackled his studies on the influence of light and other external factors on the resistance of wheat to *Erysiphe graminis* from a "metabolic resistance" point of view. This metabolic resistance is determined by the intensity of the photosynthetic, glycolytic, and respiratory activity of the host. According to him, the enabolic processes dominate over the catabolic processes in plants that are in a condition of defense, while the opposite is true for plants that are in a condition of susceptibility. Unfortunately, Sempio does not report, how far the interaction *type* can be changed by the manipulation of the environmental conditions. Therefore, it is difficult to determine how applicable to HR is his broad generalization that an unbalanced increase in the respiratory rate is always linked with a marked decrease in the metabolic defense.

The following studies have also made clear that there is a strong connection between the HR mechanism and respiratory metabolism. If potato tubers that normally interact parabiotically with certain races of *Phytophthora infestans* are treated with respiratory inhibitors such as pyrocatechol or tyrosine, the reaction will shift to the eusymbiotic type. The same effect appears after pretreatment of tubers with polyphenoloxidase inhibiting compounds (e.g., phenol urethane or potassium cyanide) or with acids from the Krebs cycle. Király and Farkas (1957) found that in all five types of wheat, of which two interacted eusymbiotically and three parabiotically with the *Puccinia graminis* race they used, the O_2 uptake was clearly higher than in the controls, after 7 to 10 days incubation. However, quantitative differences between the different reaction types were found. In eusymbiotic combinations, O_2 uptake was 6 to 12 times higher than in the parabiotic ones. On the other hand, the glycolic acid oxidase activity decreased in all interaction types, more in the parabiotic than in the eusymbiotic combinations. The decrease of glycolic acid oxidase is explained by the authors with the assumption that the prosthetic group of the enzyme (lactoflavinephosphate) is in great part withdrawn from the cell of the host by the vitamin hungry pathogen. The greater decrease of the enzyme activity in parabiotic combinations can be explained by the fact that an additional inactivation of the enzyme occurs in the dying tissues.

Another interesting attempt to gain glimpse into the dynamics of energy metabolism of eusymbiotic and parabiotic H/P complexes (rust and mildew in cereals) was undertaken by Shaw and Hawkins (1958) who pursued the decarboxylation of artificially produced indoleacetic acid (labeled with C^{14}). A sharply increasing elimination of $C^{14}O_2$, soon followed by a sharp decline, was found in both combinations.

It is impossible at this time to form a coherent—even hypothetical—opinion of the respiratory-physiological basis for HR. One thing is certain. There are basic quantitative differences underlying the eusymbiotic and parabiotic interactions. This is indicated by two facts: in the first stages of interaction *both* types show a marked increase in the respiration rate as compared to controls; also, as far as the external course of HR is concerned, the behavior of intermediary interaction types can be shifted by manipulating external conditions toward *both* sides. As would be expected, there is a sharp drop in respiration rate after a certain incubation period in parabiotic H/P combinations—contrary to the eusymbiotic ones—because the growth of the pathogen as well as metabolic activity of the host tissue stop after a short incubation period. The only thing we can say today is that in parabiotically reacting cells the intensification and the diversion of the respirator metabolism induced by the pathogen surpass the proportion of eusymbiotically reacting cells, consequently, the premature physiological breakdown of the host cell and the inhibition of the pathogen are connected with an especially high show of energy by the reacting cell of the host.

b. Photosynthesis. No special discussion is needed for the fact that assimilation activity ceases in the parabiotically reacting tissues. In this connection, studies by Gassner and Goeze (1936) show that almost no decrease of assimilation activity occurred in reaction type "i" of wheat infected with yellow rust, but a reduction of more than 50% in reaction types "0 to 2" and "4."

c. Permeability Relationships. For the maintenance of a eusymbiotic relationship there must be a physiological balance for a time between the interfaces of host and pathogen, where they are in direct contact, so that a metabolic exchange can take place between both partners. According to present knowledge, this exchange appears to occur chiefly by osmosis. According to Meyer (1940), semipermeability is maintained 6 to 14 days (19°C.) in the parenchyma of potato tubers infected with an eusymbiotic strain of *Phytophthora infestans*, while it is lost after 3 to 48 hours in parabiotic interaction. Thatcher (1942) used the rate of deplasmolysis as a measuring stick for permeability of plasma membranes and found in rust-infected wheat that permeability increases in eusymbiotic combinations; in parabiotic combinations, on the other hand, it decreases in the cells adjoining the necrotic tissue. Thatcher postulated that any factor altering the permeability of plasma membranes of a particular host tends to modify its susceptibility, provided that other changes in the host-pathogen relationship arising from the same cause are insufficient to offset this susceptibility change. This conclusion could go too far, since Thatcher's results only indicate that in parabiotic combinations a change in permeability occurs in advance of the hyphal tips, and physiological changes in advance are usual for HR. For instance, chlorotic fadings are characteristic in the tissue immediately around the area attacked by the pathogen; chlorosis is obviously due to substances that diffuse from the necrotic tissue into the adjoining cells.

Greenham and Müller (1956) used the ability of tissue to conduct electricity in potato bubers infected by *Phytophthora infestans* for quantitative determination of the damage suffered by the cells of the host in eusymbiotic and parabiotic H/P complexes. Osterhout's classic theory is that at low frequency, living tissues have low conductance, because the cell membrane is impermeable to ions but the permeability may be increased by the action of poisons. According to expectation, the conductance ability increased clearly 26 hours after infection in the parabiotic combinations, but it did not change after still another 24 hours in eusymbiotic combinations. After prolonged waiting,it increased in parabiotic combinations 25 cell diameters away from the infection focus. This "in advance effect" was missing in eusymbiotic combinations. In the latter case, the increase in conductance ability was found only at a distance of 51 cells *behind* the foremost hyphal tips.

d. Studies with Isotope Labeled Compounds. Results obtained with the use of isotope technique are summarized as follows. In susymbiotic combinations, e.g., *Phaseolus vulgaris* and *Uromyces appendiculatus, Helianthus annuus*, and *Puccinia helianthi*, and others, a strong accumulation of labeled compounds occurs ($C^{14}O_2$, $H_3P_{32}O_4$, H_2S_{35} and a series of organic compounds with labeled carbon). On the other hand, accumulation is weak or absent when labeled compounds are fed to parabiotic combinations. Since the accumulation is strongly reduced, by cutting off oxygen or by pretreatment of tissues with DPN, this phenomenon must be connected with the energy metabolism of the tissue of the host. Accumulation of isotope labeled compounds is never observed in tissues that had been attacked by necrotrophic pathogens such as *Venturia pyrina, Septoria aesculi*, and others. Shaw et. al. (1954) formulated this thought about obligate parasitism. A prerequisite for normal development of biotrophic parasites is sufficient supply of substrate to the infected tissue, from the outside. This process depends on the physiological condition, especially on the respiration of the infected tissue. If the tissue of the host is killed prematurely by the pathogen, however, it is plausible to assume that the accumulation of isotope labeled compounds stops too.

e. Quality, Position, and Age as Interfering Factors. The capacity to respond to infection with HR can be quite different in different tissues of one and the same individual. In barley plants, resistant to *Helminthosporium gramineum*, coleorrhyza tissue reacts with "necrotic pockets", while the roots "tolerate" the pathogen and are destroyed by it. The organotropically bound sensitivity is especially evident in potato and *Phytophthora infestans*. In certain carriers of gene R_1, the reaction of the leaves is to limit necrosis greatly. The parenchyma of the tubers, on the other hand, "tolerates" the invasion of the pathogen and the tubers are completely destroyed by it. The marrow parenchyma of the stem behaves in intermediary fashion. The petals—whether the plant carries gene R_1 or not—are always attacked in a normal way. Thus, there is a sensitivity gradient that is characterized by the following successive decrease in sensitivity: leaves, marrow parenchyma, tuber parenchyma and petals. The steepness of this gradient is controlled genetically and one wonders whether "minor genes" or only differentiating "effective doses" of gene R_1 are involved.

The ontogenetic variability of sensitivity is seen even in individual organs. The etoliated tuber sprouts of many *Phytophthora* resistant potato types are less sensitive at the tip than at the base. This seems generally true for meristematic tissue. On the other hand, "loci of different sensitivity" can be differentiated in the parenchyma of grown potato tubers. In parts of tissue with higher reaction rate, the pathogen stops growth relatively early.

The shift of reaction potential that appears during individual development of the plant and its organs, might be closely related to those organoor histotropically bound sensitivity differences. In half-grown wheat plants, for instance, the leaves react differently to infection with *Puccinia triticina*, depending on their position on the stalk. Sensitivity decreases toward the base.

The shift of reaction position with progressive development of the plant in rust diseases of cereals is of special interest. Although the reported findings do not offer a completely uniform picture, it seems that sensitivity usually increases during individual development of the plant. It has also been established that in virus diseases sensitivity increases with progressive development of the plant (*Salanum melongena* and tobacco mosaic virus). However, the opposite tendency is also observed, for instance in *Phaseolus vulgaris* and *uromyces phaseoli*.

B. The Influence of External Factors on the Course of the Hypersensitivity Reaction

The effect of external factors on HR may be a direct or an indirect one. In the case of the latter, predisposition is changed. Literature, accumulated since Ward's (1902b) time—chiefly in the field of cereal rust research—is so abundant that the following references have to suffice:

1. *Effect of Temperature*. The speed with which HR expires as in every living process is largely determined by temperature. Also the degree of sensitivity and the clinical result of interaction usually depend on temperature. Thus, sensitivity increases in wheats and *Puccinia glumarum* as the temperature increases; however, in wheats and *Puccinia graminis* sensitivity decreases. Johnson studied the behavior of a whole series of physiological races of *P. graminis* on standard wheat types and found that the proportion of temperature influence is co-determined by the individual H/P complex. Intermediary interaction types usually show the greatest liability. Among others, Straib (1940), showed for yellow rust and Hayden (1956) for black rust, that the age of the plant has a bearing here too.

Similar relationships have been obtained in other diseases elicited by fungal pathogens. Thus, the mildew-resistant wheat type "Hope", does not give the defense necrosis reaction at temperatures of 24° and 28°C that is, it is susceptible at these temperatures. The reaction of *Phytophthora* susceptible potato tubers is shifted toward the parabiotic side by lowering the temperature below the 10°C limit.

In virus diseases the temperature also exerts a decisive influence on the interaction between host and parasite, in many cases. The interaction between *Nicotiana glutinosa* and tobacco mosaic virus should be mentioned, as an exemplary case. In temperatures below 28°C the plants react with "local lesions;" under higher temperatures (35°C) with systemic chlorosis. According to Holmes (1932), local lesions are much smaller at 16 to 18°C. than at 20 to 25°C At 10°C lesions do not develop.

The predisposition of the plant is also influenced by temperature. According to Gradinaroff (1943), under prolonged sublethal pretreatment at temperatures above 35°C, the potato tuber loses its ability to localize infections with the otherwise avirulent *Fusarium* species. Straib and Noll (1944) found that in wheat varieties more or less resistant to *Puccinia triticina* sensitivity is reduced by pretreatment at 50°C for 60 seconds. Thus, in the highly resistant Malakoff varieties, the reaction type changed from "i" to "2 to 3." However, the leaves regained their high sensitivity if the plants were returned to normal temperatures for 4 days. Similar results were obtained in pods of *Phaseolus vulgaris* which had been treated with 44°C for 2 hours and inoculated with *Sclerotinia fructicola* or *Botrytis cinerea*.

In virus diseases, too, a gradual decrease of the sensitivity after treatment with supramaximal temperatures, occurs now and then. Thus, if leaves of *Nicotiana glutinosa* are exposed for 40 seconds to 50°C, and then infected with tobacco mosaic, the diameter of lesions, becomes about 80% greater than in controls, according to Yarwood (1958); on the other hand, the time for lesions to develop decreases.

2. *Effect of Light and Atmospheric* CO_2 *Content.* In cereal rusts, as a rule, optimal light supply promotes development of the pathogen. At any rate, available literature does not always make it clear whether inhibition of the pathogen under inadequate lighting is due to increase of sensitivity, or to a "hunger crisis" of the fungus due to insufficient nutrition. Bewer (1934) varied the day length in barley and *Puccinia glumarum* under average temperature of 9°C. With 15-hourly or continuous exposure to light a highly susceptible variety, "Pasier", showed reaction type "0" estimated only after lack of sporulation of the pathogen. In using lettuce plants the stems of which reacted with necrotic lesions to individual *Pellicularia* strains, Flentje (1957) found that the hypersensitive reaction of the host was lessened by exposure to reduce light intensity. On the other hand, Hassebrauk (1940) succeeded in shifting the reaction type toward increased sensitivity in wheats and *Puccinia triticina* by keeping the plants for two days in darkness during the incubation period. The proportion of induced changes and the time of treatment at which the response was greatest, was different in the different H/P combinations.

Our information on the influence of CO_2 content of the air is limited to the rust diseases. Gassner and Straib are responsible for most detailed studies of this problem. Insufficient CO_2 supply primarily lowers the frequency of infection in yellow rust of wheat, but does not shift the degree of sensitivity. In concentrations above 1.5% CO_2 in the air, changes do occur in the infected host tissues that resemble the rust type 0. "Necrotic discoloration" increases considerably as the CO_2 content increases further; borderline concentrations, where the eruption of pustules in brown rust of wheat is prevented, are not equally high in different interaction types, but are lower in intermediary interacting H/P combinations than in the susceptible ones.

3. *Influence of Mineral Salt and Water Supply.* Because there was hope of obtaining with nutritional-physiological experimentation an insight into the mechanism that prevents normal development of the pathogen in the hypersensitive tissues, special attention was given to the relationships between nutrition of the plant and its immunological behavior. Since this point will be treated in more detail elsewhere, the following remarks should suffice. As a rule, only hosts with intermediary sensitivity respond to differences in mineral salt supply. Generally speaking—again in cereal rusts—it was found that lack of nitrogen increases sensitivity, surplus of nitrogen decreases it. Potassium exerts an antagonistic influence. The significance of phosphorus could not be unequivocally clarified. Obviously it depends on the simultaneous supply of nitrogen and potassium. Development of obligate biotrophic parasites is better if the host's growth is greater. Unfortunately, many studies do not differentiate sharply enough between the reaction type and the frequency of infection per unit area of the surfaces exposed to the parasite. The many discrepancies in the literature can be explained from the use of extreme eusymbiotic or parabiotic H/P combinations. This is especially true in cases where the effect of mineral salt nutrition was studied on the reaction to diseases elicited by favultative biotrophic or necrotrophic pathogens (e.g., potato and *Phytophthora infestans*).

Chessin and Scott (1955) found that quantitative differences in the size of local lesions in *Nicotiana glutinosa* infected by tobacco mosaic virus depend on mineral salt supply. Lack of iron or sulfur causes an increase in size of local lesions. This does not happen in calcium or magnesium deficient plants.

In a bacterial disease (*Zea mays* and *Phytomonas stewarti*) seedlings insufficiently supplied with nitrogen reportedly showed "small necrotic lesions but little or no wilting of the invaded leaves." This relationship confirms the opinion that in bacterial diseases, too, interactions appear to set in that can be compared to the hypersensitivity reactions so often observed in mycoses.

As far as water supply of the plant is concerned, available information suggests that a low water content increases the sensitivity of tissues. If the plants that have become hypersensitive through wilting are brought back into normal conditions, they resume a normal behavior.

4. *Effect of Narcotics*. Stakman (1914) was the first to work on the influence of gaseous narcotics on degree of sensitivity. He found that treatment of plants with chloroform made an immune plant somewhat susceptible to rust. Gassner and Hassebrauk (1938) confirmed Stakman's results in a whole series of cereal rusts. In certain combinations they succeeded in changing the reaction type from 0 to 4. In potato and *Phytophthora infestans*, treatment of hypersensitive tubers with sublethal concentrations of alcohol provoked a decrease of reaction rate; simultaneously the pathogen was able to penetrate deeper into the tuber parechyma and fructify. Tomiyama et. al. (1957) who confirmed these findings assume that the effect of narcosis to decrease sensitivity is connected with the respiration (dehydrogenase) system of the host.

5. *Effect of Chemotherapeutants*. Gassner and Hassebrauk (1936) were the first to increase sensitivity and with it resistance of cereals toward rust with the help of a systemic "fungicide", by temporarily transferring young plants into nutrient solutions containing sulfides. The reaction type was thus lowered from 3 or 4 to "0" and "1." The same effect was obtained by Hassebrauk (1951) with sulfonamides and sulfones, which he mixed into the soil. He explains the effect of these compounds as a separation of the structurally similar and for some fungi essential *p*-aminobenzoic acid from its protein carrier, just as in human medicine. Hotson (1953), reported still other organic compounds that are effective against stem rust. In accordance with Hassebrauk's assumption, Hotson reversed the rust inhibiting effect of sulfadiazine by using *p*-aminobenzoic acid.

According to Sempio (1912), cadmium—used as $Cd\ (NO_3)_2$—acts as a systemic fungicide against wheat mildew. Meyer (1951) explains the effect of cadmium as an increase in sensitivity of epidermal cells of the host toward the pathogen.

Also, antibiotics applied systemically to the plant can increase sensitivity, at least as far as external symptoms are concerned. The effect of streptomycin in potato and tomato against *Phytophthora infestans* is explained by Vörös et. al. (1957) as a streptomycin induced increase of polyphenolase activity. But "the antibiotic was shown to be totally inactive when tried directly as a 'substrate' in the assay of polyphenolase activity", thus, the streptomycin effect must be an indirect one.

6. *Effect of Irradiation*. Norell (1954) showed, that UV-irradiation lowers the resistance of potato tubers to *Fusarium* species, whereby the marrow parenchyma responds better to treatment than the peripheral tissues. X-ray treatment of young wheat and oat plants decreases the sensitivity of intermediary interacting H/P complexes to *Puccinia graminis*; it is unsuccessful in extremely susceptible or resistant hosts. In postinfection treatment, on the other hand, resistance of flax against *Melampsora lini* increases; but only seemingly, because as Schwinghamer (1957) showed, the increase in resistance is based on direct damage to the pathogen by irradiation.

In retrospect, it can be said that the mechanism of HR, that is primarily determined by hereditary factors in the host as well as in the pathogen is subject, within certain limits, to changes in external conditions and to those of ontogenetic nature. The influence that temperature exerts on the course of HR is especially informative. It not only determines the rate of interaction but also the "biological balance" between the interacting partners. Furthermore, we have seen that the living cell can "repair" inactivation of the HR mechanism caused by supramaximal temperatures. This shows that the ability of the host to respond to the attack of a pathogen with HR is inseparably bound with a certain physiological condition of the cell of the host. If we use the term "*Fliessgleichgewicht*"* borrowed from biochemistry, and correspondingly presume that the anabolic and catabolic processes are in equilibrium in the uninfected cell, HR can be interpreted as the result of a deviation of this equilibrium induced by the pathogen. This deviation would then be the cause for the premature death of both interacting partners. At any rate, this does not mean much as long as we don't know more about the mode of action of metabolites of the pathogen that induce

* Identical with "stationary state."

HR, and about the nature of the factor that prevents further growth of the pathogen in the hypersensitive tissues.

V. HYPERSENSITIVITY REACTION AS AN IMMUNOLOGICAL PROBLEM

The main problem here is as follows: What is the nature of the factor that brings about early death in the development of the pathogen in the parabiotic H/P complex? This question was raised earlier by Ward (1902). Although he was unable to prove the existence of such a factor, he summarized his concept, mainly based on histological findings, as follows: The antagonism between host and pathogen "must be due to something far more subtile than a mere soluble poison oozing from the cells."

Since Ward's day, this question has been discussed in many works and opinions about the antipathogenic principles differ widely. According to many authors, nutritional physiological factors which determine the suitability of host tissue as nutritional substrate for the pathogen, govern whether a typical pathogenic relationship occurs or not. Other authors say that inhibition of the pathogen is due to preformed inhibitors. Since the finely adjusted relationship between pathogen and host, e.g., rust diseases, can be explained by neither supposition, not so much the absolute content of specific nutritional products or specific inhibitors, but the relationship between these two has been suggested as being involved in the establishment of a pathogenic relationship. Other authors have concluded, mainly on the basis of histological findings, that postinfection chemical changes of the host tissue hinder the normal development of the pathogen. Finally, to reconcile the two seemingly contradictory hypotheses it has been assumed that some substances diffuse from the pathogen that is dying because of lack of nutrition, into the cell of the host and its surrounding; these substances elicit the physiological breakdown of the host cell and thus cut the parasite off from its sources of nutrition.

A. Lack of Nutritional or Growth Factors as Antipathogenic Principle

This assumption stems from the thought that corresponding to the specialization of the pathogen on its host, there is a notable specificity in the nutritional requirements of the pathogen as well as in the chemical make up of the host. Hereby, all H/P combinations involving necrotrophic pathogens are *a priori* excluded from consideration.

The simplest way of investigating this thesis is to compare the pressed out juices of closely related hyper- and normosensitive hosts with respect to their adequacy as nutritional substrates for the given pathogen. All attempts of this kind have failed until now. Yet, many authors still adhere to the "nutritional hypothesis." The most important arguments for this are: (1) lack of success in all attempts to find an inhibitor in the hypersensitive host that is absent in the normosensitive one—irrespective of whether it is preformed or is formed after infection; and (2) the obligate biotrophic character of the pathogen in question.

The structure of the specific proteins of the host, and their precursors, as well as of the carbohydrates have been invoked to explain the varied behavior of the host plant toward a certain pathogen or one of its races. Leach (1919) and Wellensiek (1927) were determined followers of this hypothesis. Fischer and Gäumann (1929) used it to explain the extreme specialization of physiologic races of a rust fungus for specific host types. The specific structure of proteins was most often used as an example.

Gassner and Franke (1934) also felt this way; they found that in cereal rusts, the infection type increases with the protein content, under differentiated potassium and abundant nitrogen supply. Since, however, no correlation could be proven between varietal resistance and protein values, they postulated that qualitative differences within the existing protein compounds are coordinated with the hypersensitivity connected with the individual H/P complex. Finally, Johnson's (1953) physiological interpretation of the action of the genes that control HR in cereal and flax rusts is also based on a purely nutritional concept. According to him, genes that control HR are "enzyme producers" and an individual rust race could function as a pathogen only when it finds the enzymatic system needed for successful parasitism in the host. This purely hypothetical concept, however, does not agree with the interpretation according to which the production of a certain enzyme is bound directly to the dominant allele of the gene.

Attempts have also been made to find correlations between carbohydrate content (mainly of sugars) and varietal differences in hypersensitivity. These attempts can be traced to Comes (1913) who claimed to have found a positive correlation between sugar content and susceptibility. Even with the use of modern methods for separation of various types of sugar (glucose, fructose, and maltose) no difference could be found in the sugar content among various reaction types of wheats and *Puccinia triticina*.

B. Preformed Inhibitors as Antipathogenic Principles

It is quite natural that the demonstration of the preformed inhibitors as a cause of inhibition of pathogens in hypersensitive tissue can only be considered successful when the following presumptions are fulfilled: (1) There should be no possibility that inhibitors separated from the host tissue are preoperative artifacts; (2) it should be proven that the *in vivo* concentration of the factor in question is sufficient to bring about death of the pathogen in a short time; (3) there should be a positive correlation between the *in vivo* concentration and the sensitivity degree that the host exhibits toward the given pathogen.

Since it has been impossible yet to show that the preformed inhibitors meet these requirements the following condensed survey of studies done on this problem should be sufficient.

1. *Acid Content and Hydrogen Ion Concentration of the Host Tissue*. Comes (1913) was the first to suggest that acidity of the cell juice plays an important role in the resistance of the plant to its potential pathogen. His results, however, could not be confirmed by later authors—not even when hydrogen ion concentration was used as measure of acidity.

Hassebrauk and Kaul (1957) studied the content of ascorbic, citric, oxalic, and malic acids in extracted juices of wheat species that behave differently toward *Puccinia triticina*. No connections could be proven between acid content and resistance. In wheats and *Puccinia graminis*, Pilgrim and Futrell (1957) achieved the same result for ascorbic acid.

2. *Content of Phenolic Compounds in Tissue of the Host*. The thought that preformed toxins, particularly of phenolic compounds, could cause inhibition of the parasite in hypersensitive tissues, found many followers, since Angell et. al. (1930) proved in *Allium cepa* and *Colletotrium circinans* that the relative resistance of brown skinned bulbs is due to the antibiotic effect of phenolic compounds (protocatechuic acid, catechol) that diffuse from the dead exterior scales into the surrounding area and thus protect the inner scales from attack by the pathogen. Dufrenoy's findings according to which there is an accumulation of phenolic compounds in infected tissues, also encouraged further search for preformed substances that have a toxic effect on fungal and bacterial pathogens. But the result of these attempts was disappointing. Although consider-

able quantities of phenolic compounds could be found in the living tissues, no significant differences could be obtained between the phenol content of hyper- and normosensitive hosts. Only in the case of wheat varieties resistant to *Puccinia triticina* did the pressed out juice have a higher content of protocatechuic acid than the susceptible ones.

Notwithstanding these discouraging results, many authors still maintain that preformed phenolic compounds somehow have a causal connection with HR. Newton and Anderson (1929) and Scott et. al. (1957) suppose that in a parabiotic H/P complex activation of the phenolic compounds occurs after the infection. In eusymbiotic H/P complexes semi-permeability of cells is maintained for some time, the phenolic compounds are therefore held back in the cell of the host and thus cannot be effective; in parabiotic combinations, on the other hand, they are "set free" by the dying cells and they can then develop their antipathogenic effect. This hypothesis actually presumes that the phenolic substances found *in vitro* are also present *in vivo*, in concentrations that are sufficient to stop the growth of the pathogen. No quantitative analyses on this are available as yet.

3. *Enzymatic Activity of the Host Tissue and Hypersensitivity*. Quite a number of workers tried to establish a correlation between enzymatic "equipment" and immunological behavior of the plant. Grechushnikov (1939), for instance, claimed to have found in potato varieties resistant to *Phytophthora infestans* that the activity of peroxidase is higher than in susceptible ones. In using potato varieties resistant due to hypersensitivity, Kammermann (1951) failed to find such a relation. Hassebrauk and Kaul (1957) determined the content of cytochrome oxidase, ascorbic acid oxidase, polyphenolase, and peroxidase in wheat varieties hypersensitive to as many races of *Puccinia graminis, P. triticina*, and *P. glumarum* as possible. "Michigan Amber", highly susceptible to all three types of rusts, was used as a control. Moreover, plant of highly resistant varieties were treated with enzyme poisons and compared with the untreated plants for fermentation activity. The cytochrome oxidase activity was found to be smaller in eusymbiotic H/P complexes than in the parabiotic ones; the opposite was true for the behavior of ascorbic acid oxidase and polyphenolase activity toward *Puccinia triticina*. The authors considered that plants reacting in a more or less hypersensitive fashion are characterized by a "quicker metabolism" and that the content of ascorbic acid and phenols in the host plant is probably involved in determining the infection type.

C. Hypersensitivity Reaction as "Defense Mechanism" Sensu Stricto

The possibility last mentioned, leads to theories according to which a defense reaction is initiated by infection which limits further growth of the pathogen in hypersensitive tissues. There are two possibilities, *a priori:* the pathogen is deprived of nutritional supplies by the above mentioned histological changes, or the activation or new buildup of one or more toxic principles is produced by the infection, and thus further penetration of the pathogen is stopped.

1. *Postinfection Interruption of Nutritional Supply to the Pathogen as Antipathogenic Principle?* The former possibility was considered very early. It is naturally only applicable to obligate biotrophic parasites and presupposes that (1) death, or at least damage, of the host cell occurs before death of the pathogen, and (2) the pathogen releases some substances into the cell of the host that have a lethal effect on the cells of the hypersensitive host. The former supposition can be proven. Conclusive evidence for the premise mentioned second is lacking; the proof of specific toxins released by the pathogen into the cell of the host, or the *specific* reaction of the host to these, cannot be proven as yet.

Another hypothesis that also presupposes the maintenance of a living state for the establishment of a normal pathogen relationship, comes from Gassner and Hassebrauk (1938). According to this hypothesis, eusymbiotic interactions between the cell of the host and the pathogen are based on a quantitative relationship between the toxins eliminated by the pathogen into the cell of the host and the neutralizing "antitoxins" of the cell of the hot. Parabiotic interactions should consequently be ascribed to the inability of the host to mobilize antitoxins with sufficient speed. In establishing this hypothesis, Gassner and Hassebrauk relied on the experience that the sensitivity is lowered when the plant is pretreated with narcotics, whereas the nitrogen content of the tissue increases. Contrary to the concept prevailing in animal pathology, the ability of antitoxin production would be correlated with susceptibility to the disease producing agent.

2. *Postinfection "Antibodies" as Antipathogenic Factors.* A short survey of literature about HR will show that many workers have counted on the possibility that during the interaction substances are formed that have a toxic influence on the pathogen. A whole line of indications have corroborated this concept. Bernard (1990) found that embryos of

Loroglossum hircinum, an orchid, are protected from infection with a virulent strain of *Rhizoctonia repens* by preinfection with a less "aggressive" representative of this fungus. The protection was restricted to the preinfected tissue. Similar attempts by Müller and Börger (1940) left barely any doubt that the hypersensitive cell of the host is actively involved in HR as a producer of defense factors. The way to a more dynamic interpretation of HR was paved, since during the last few years it was proven that the plant can really produce "defense bodies" in response to fungal attack.

a. "A Priori" Indications. In obligate biotrophic pathogen (viruses included) the scientist always faces the problem as to whether the postinfection inhibition of the intruder is only caused by the premature death of the infected cell. This difficulty does not exist in the case of facultative biotrophic or necrotrophic pathogens. It can be explained herewith that important progress in causal-analytical clarification of the hypersensitivity problem was seen only after the general occurrence of HR was recognized in H/P combinations too, in which no obligate biotrophic organism participates. *Venturia or Colletorichum*, can be easily cultured on many inert nutritional substrates and thus prosper better on such substrates than on the natural host. In these cases one cannot understand why the pathogen, after initial normal development in the hypersensitive tissue, should stop growth because of lack of nutrition. On the other hand, the contention that the "demarcation tissues" which often developed postinfectiously hinder further advance of the pathogen, lost its importance when it was recognized that inhibition of the pathogen can be observed even before the histological differentiation of the demarcation tissue takes place. The ultimate contention that the pathogen stops growth possibly because of "selfpoisoning" (i.e., elimination of metabolites that are poisonous for the pathogen itself) cannot readily be discarded. But it is necessary, in this case too, to postulate as far as the behavior of the pathogen in a normosensitive host is concerned, that the effusion of "self-inhibitory" substances depends on the chemical composition of the tissue of the host.

b. Experimental Indications. It was proven with potato and *Phytophthora infestans* that the tissue of the host is transformed during parabiotic interaction into an environment that is inhospitable to the pathogen: If the parenchymal tissue of tubers of a R_1 gene carrier is inoculated with a parabiotic race of the fungus, the tissue loses its property of serving as a host tissue even for those races of the fungus that would otherwise develop a high virulence on it. Even with simultaneous inoculation of parabiotic and eusymbiotic races, the eusymbiotic partner does not develop normally. This result shows that which the host tissue is transformed during interaction. Since both races can grow normally one next to the other in a narrow space in tubers that are susceptible to them both, this "inhospitable" condition—that is brought on the resistant tuber by preloading with a parabiotic *Phytophthora* race—cannot be based on an antagonism between the two races. Thus, this must be necessarily due to an interaction in which the host is equally involved with the pathogen that interacts with it parabiotically.

The necrotic tissue does not accept other organisms that grow on living potato tubers (e.g., *Fusarium caeruleum*) or on tuber slices killed by heat. Thus, lack of nutritional substances cannot be considered the cause for inhibition of fungal growth in that parabiotically reacting tissue. But no total immunization of the tuber takes place. The eusymbiotic race can develop normally outside of the "vaccinated" tissue area and its immediate surrounding, directly attacked by the parabiotic race.

From the foregoing the fact that there is a close correlation between reaction rate and resistance degree, we may conclude that a principle is activated or newly formed in the parabiotically interacting tissues that has a toxic effect on the microorganisms. This principle could be an "antibody" in the original sense of the word. But, since the animal pathologist associates the concepts of specificity and humoral immunity with this term, the postulated antibiotic principle is called "phytoalexin." This term refers only to the function as "antibody"; it does not include an statement about the chemical structure of the active principle.

Arnaudi (1942), studying a problem similar to the one of Müller and Börger, soon afterward established that in the same H/P combination, injection of potato tubers with parabiotic races of the causative agent gives a local protection against attack by eusymbiotic races.

Johnston's and Huffman's (1958) experiments explored the existence of an antagonistic effect of rust fungi *in vivo*. Their study points in the same direction: if wheat leaves are covered with spores of *Puccinia coronata*, a pathogen of oats, the germ tubes of the fungus penetrate into the host tissue but only elitic "heavy flecking" there. If the leaves are infected two days later with a virulent race of *Puccinia triticina*, the frequency of microscopically recognizable *triticina* infections is strongly reduced; on the other hand, only necrosis occurs in the areas preinfected with *P. coronata* (reaction type 1). Outside of the preinoculated area, reaction is normal (reaction type 4). The writers leave open the question whether an effect of substances "produced by the latent mycelium of an organism (i.e., *Puccinia coronata*) that was not able to establish itself parasitically", is involved or only a purely mechanical blocking of the stomata through preinfection. Neither might occur, according to the results obtained with experiments on potatoes. Most probably the inhibition of *Puccinia triticina* in the mesophyll is due to changes in chemical constituents after infection with the parabiotic partner.

c. Demonstration of Interaction Products with Antibiotic Effect. Demonstration of such substances succeeded under adequate experimental conditions in combinations, *Phaseolus vulgaris* (inner epidermis of seed cavities) and *Phytophthora infestans*, and *Ph. vulgaris* and *Sclerotinia fructicola*, that interact "in a hypersensitive fashion." In three other hosts (*Capsicum annuum, Pisum sativum, Vicia faba*), the same result was achieved after infection with the above mentioned pathogens. It was also shown that development of the antibiotic principle occurs a few hours after infection in parabiotically reacting tissues. Under the experimental conditions used, the effective principle is developed and diffuses from the site of interaction in quantities a thousand times greater than necessary for preventing fungal growth at the focus of infection.

The specificity of the effective principle (phytoalexin) is small. For instance, it is effective against *Uromyces trifolii* or *Colletotrichum lindemuthianum*. It passes through semipermeable membranes (e.g., Cellophane) which points to a relatively small molecular weight; it has properties that characterize it as a "hydrophilic lipid." It is effective within a pH range of 4.0 to 7.5. Its antibiotic effectiveness is not influenced by the presence of substances that could serve *in vivo* as nutritional substances for the pathogen. The output capacity increases with increasing age of the host tissue. This recalls the fact that older plants are usually more sensitive than young plans ("age resistance"). The phytoalexin is mostly adsorbed by the parasitized cells and surrounding tissue. This might be especially conclusive for the understanding of the HR mechanism. Because, if the phytoalexin is not fixed, larger and larger quantities would flow into the surrounding tissue, and the concentration of phytoalexin critical for the pathogen would be attained much later or not at all at the focus of infection.

As mentioned previously, short-term treatment of tissue of the host with high, but not lethal, temperatures leads to a temporary blocking of

HR. As expected, it was found that the temporarily "desensitized" tissue are incapable of responding to the infection with production of phytoalexin. When, however, the host tissue regains its original sensitivity, phytoalexin elimination also becomes normal.

The question is whether those results can be applied to H/P combinations when an obligate biotrophic parasite is involved. Even through much seems in favor of this concept, a generalization of findings obtained by Müller in "unnatural" H/P combinations is not yet permissible. In the case of obligate parasitism, one should take into account the possibility that specific nutritional or growth substances, essential for the life of the parasite, may be of greater significance for the initiation of HR.

V. PROBLEM OF "ACQUIRED IMMUNITY"

The concept of "acquired immunity" is, like that of hypersensitivity, borrowed from human pathology. It is based on the experience that a host once cured of disease has a more or less prolonged defensive protection against a second infection. Contrary to this "acquired" immunity, is considered "inherent" ("natural" or "congenital") immunity, characterized by the fact that it is inherent in the organism from the beginning of its development. When it was proven in animals at the turn of this century, that the "condition" of acquired immunity is associated with the existence of specific circulating "antibodies" in the blood, attempts were made to prove a similar defense mechanism in plants. Based on experiences in mammals, attempts were made to reach that goal in two ways. First, to demonstrate that the resistance of the plant is increased by preinfection with "attenuated" strains or by "vaccination" with metabolic products of the homologous causative agent. Second, lysin-, precipitin- or agglutinin-like bodies that protect the plant from attack of a potential parasite were searched for. Of secondary significance was the question whether performed or postinfectionally obtained specific "antibodies", are involved as in animal pathology.

In the following attempt to summarize the result of these efforts, all resistance changes should be excluded that can be obtained by systemic application of chemotherapeutics, e.g., sulfonamides or antibiotics, or by manipulation of external circumstances. By definition they do not belong into the category of "acquired" immunity. On the other hand, this concept should not be taken too literally. Thus, it would not be right to exclude from the following considerations all cases in which immunization of the whole plant fails to be evident. This is even less indicated because research in animal pathology only recently began to pay more attention to the phenomenon of locally acquired immunity and to material changes of the host tissue that are limited to the infection focus. Also, the existence of acquired immunity should not be linked with the postinfection appearance of specific "antibodies." In animals, they only make the pathogen more susceptible toward the resistance mechanism linked to the diseased tissue. The interpretation of animal pathologists predominantly tends toward the conclusion that the tissue defense reactions of the immune and non-immunized animal are qualitatively similar and that after acquisition of the immune state the existing reactions appear more rapidly and heightened with greater intensity. No new reactions appear.

No survey of the subject acquired immunity in plants can be given with the framework of this chapter. Chester's (1933) and Vavilow's (1935) summary of the problem and the literature cited by Gäumann (1951) and Hess (1949) provide a basic bibliography. A few strategically important papers are reported as follows.

The first attempts to prove the existence of "acquired immunity" within the plant kingdom stem from the French scientists Beauverie (1901) and Ray (1901). They reported success in obtaining an increased resistance in their experimental plants through inoculation of the soil with, or by direct application of, avirulent strains of the disease-producing agents in question.* However, the interpretation of these experimental findings as a proposed working hypothesis was accepted with considerable skepticism by many phytopathologist. This was due, on the one hand, to lack of precision of the published results, and on the other to the fact that the results of both authors as well as of later authors could not be corroborated. Finally, it was simply denied, from purely theoretical considerations that the plant has the ability to "acquire" resistance. The main arguments were lack of a circulation system corresponding to blood circulation of animals and the unsuccessful attempts of other scientists to prove the presence of antibodies in resistant or immunized plants that could be compared to the classic antibodies of animals.

If we exclude from our observations all studies that do not have unequivocal results, or whose results could not be confirmed by later authors, the general interpretation points to the fact that plants cannot respond to infection with constitutional changes that project the whole

*Other "vaccines" too, e.g., killed cultures or mcelial extracts, were used by these and later authors.

body against a second infection. Although it was established that a reduction of attack rate or severity of disease can be effected by pretreatment of germinating seeds or young plants with culture filtrates or other "caccines", it was mostly not ascertained whether the immunization effect was caused by changes in the reaction potential or purely by a decrease in infection rate per unit of plant surface exposed to the pathogen. The results obtained by Zoja (1925) and Hess (1949) in barley or wheat the *Helminthosporium sativum* are very conclusive: both authors were able to achieve a clear reduction of disease attack in young plants that have been transplanted into the open after the seeds were cultured on a substrate to which ground mycelium of the disease-producing agent was added as a *vaccine*. However, Hess showed that the young plants suffer greatly under vaccination (poor development of roots, slower growth in comparison to controls); also, the same immunological effect could be achieved when young plants are cultured in highly concentrated nutritional solutions to which no ground *Helminthosporium* mycelium was added. Hess, to whom we are indebted for the most critical studies of this problem leaves the question open whether the "vaccination results" obtained by him and Zoja can be actually considered an immunization of plants in the true sense of the word. They could also be the result of an unspecific displacement of the reaction potential that corresponds to what we can, for instance, observe after treatment of plants with antibiotics, etc.

Where resistance is increased locally, the situation is clearer. As previously mentioned, the attacked tissue becomes more inhospitable for the parasite in parabiotic H/P complexes. The range of inhospitality produces by the primary infection is most significant. However, Bernard's (1909) studies in *Loroglossum* seedlings and *Rhizoctonia repens* and *Rh. lanuginosa*, as well as the studies of Müller ad Börger (1940) in potato tubers and *Phytophthora infestans* showed that the range does not exceed a relatively limited number of cell layers around the necrotic tissue. On the other hand, Gäumann et. al. (1950) demonstrated the antibiotically effective substance, produced by *Rhizoctonia repens* metabolites in orchid bulbs, even at a distance of 20 mm. from the tissue complex, that had been under direct influence of the parasite's metabolites. This points to the fact that some secretion product of *Rh. repens* diffuses further into the tissue of the host and induces the formation of and antibiotically effective substance.

The *Pyrus communis* and *Viscum album* case should be mentioned as an outstanding example of "sensitization at a distance." In the pear species "William Christ", that is characterized by an inherent lability, sensitivity of the trees increases under consecutive yearly infections with mistletoe; this means that the tendency to react to new infections with necrosis and to cut off the penetrating mistletoe seedlings from the remaining tissue rises. But this sensitization goes no further than the infected branch of the tree. Thus, here too, in this extreme case, the host cannot acquire immunity as a whole.

This evident inability of the plant to acquire "humoral" immunity, prompted many authors to feel that immunological effectiveness of the plant as limited to the "inherent" immunity types. This thesis, however, can be applied only to cases where freedom from attack would depend on the presence of preformed resistance factors. In the case of resistance due to hypersensitivity the thesis misses the point. In this case immunity itself is basically not "inherent." Only the reaction potential is inherent. This enables the plant to localize the disease-producing agent at the infection focus. In other words: only the ability to achieve resistance at the infection site, and at this site alone, is inherited, not the resistant state itself. The state of resistance is acquired only when the plant comes in contact with the causative agent. Then the mechanism becomes operative which changes the tissue parts, attacked by the parasite, from in "indifferent" into a "resistant" state.

VI. EPILOGUE

An attempt was made above to survey the present status of our knowledge on the essence and mechanism of HR. Such an attempt necessarily must leave a feeling of dissatisfaction behind it. This is true because the accumulated experiences are not sufficient to correlate them logically with an acceptable theory of hypersensitivity. But our knowledge is sufficient to present a hypothetical picture that brings us closer to an understanding of what we consider to be a hypersensitivity reaction. Let us proceed from the following arguments:

1. HR is a process in which the pathogen is just as actively involved as the host.

2. As in every living process, the course and clinical result of HR is controlled by a whole series of various factors.

3. As already recognized by Ward, the interaction ending in the death of both partners cannot be explained on a completely nutritional-physiological basis.

4. On the other hand, it has been possible to show, under greatly simplified experimental conditions, that a substance toxic for the pathogen is eliminated after infection.

5. The formation of this substance is correlated, in space and time, with the physiological breakdown of the cell of the host *as well as* of the pathogen.

6. The HR system can be inactivated temporarily by manipulation of external circumstances; simultaneously, the tissue loses its ability to respond to infection with production of an antipathogenic substance.

These findings do not leave any doubt that the postinfection formation of such a substance, at least in cases studied up to now, is the direct reason for the antipathogenic character of HR. Thus, HR can be considered a defense reaction *sensu stricto*.

Whether we call this defensive principle antibody or something else, is a question of terminology. The writer prefers the term phytoalexin, so that confusion with the classic antibodies of animals is avoided. Such a terminological demarcation is even more strongly indicated since phytoalexin is a defense principle directly aimed against the pathogen.

What then is the nature of the system that we can hold responsible for the postinfection formation of phytoalexins? We have to rely here mainly on suppositions. But much evidence points to the fact that HR is an enzymatic interaction in which reactants are involved that the supplied by the cell of the host as well as by the pathogen. The fact that the mechanism underlying HR is very sensitive to narcotics and enzyme inhibitors is a strong argument in favor of this concept. Also genetic arguments favor such a concept. According to the present concept, the genes—represented by their dominant alleles—have an enzyme-producing function. Our hypothesis agrees with this idea because the genes that control resistance of the host or avirulence of the pathogen have been proven mostly to be dominant. Consequently, the gene-for-gene system drafted by Flor (1955) and others and expressed in physiological terms represents a reactant-for-reactant system.

At first, this conclusion seems to disagree with the fact that antibiotic effectiveness of the heretofore proven phytoalexins is nonspecific. This only appears to be a dilemma. Actually, the specific alternative is not hypersensitivity, but normosensitivity. This is seen clearly when we transfer parasites such as *Phytophthora infestans* into "wrong host plants." Then it is found that "hypersensitivity" not "normosensitivity" represent the norm. The fact that normosensitive behavior is considered the rule is purely because we recognize the existence of a pathogen only when it goes astray onto a normosensitive host. The unsuccessful attempts of the pathogen to take hold on a hypersensitive host plant escape our notice under natural conditions.

Resistance due to hypersensitivity was explained differently as the result of a phylogenetic adjustment process of the host plant to its surrounding parasitic flora. This thought presupposes that the respective plant type or its phylogenetic ancestors were in direct contact with the potential disease-producing agent for a longer period of time. Upon closer examination, such a theory proves faulty, at least within this context. Thus, plants that can be found naturally only in the arid climate areas of Australia (e.g., *Erythrina vespertilia* Benth, *Clianthus formosus* Ford et. Vick. and *Cassia sturtii* R.Br.), proved to be hypersensitive to *Phytophthora infestans* , a pathogen restricted to humid climates. In this case, the ability to respond to the attack of the pathogen with HR, can not possibly be the result of an "adaptation process." Another important argument; the center of *Phytophthora infestans* should be searched for, according to the latest studies, in the southern part of the North American continent. Hypersensitivity should thus undoubtedly be considered a phylogenetically original situation. In the process of mutual adjustment above all only the parasite is the active (i.e., through mutation and recombination), the adapting partner; the host has only the passive mole of a selection sieve. On the other hand, once the pathogen has taken hold on the new, normosensitive, host, a second selective process counteracting the first one begins in which the hypersensitivity of the host plant represents the positive selection mark. Of course, as unfortunately the experiences of plant breeders have shown, in the long run the pathogen is always the superior partner because of its greater multiplying frequency as compared to that of the host.

The mechanism underlying HR deserves the interest of the biologists even beyond the purely pathological. Why, we may ask, are so may microorganisms incapable of utilizing and destroying "tender" tissue, such as that of a carrot root, even though they thrive richly on al sorts of artificial nutritional media and are characterized by low susceptibility to toxic substances of a higher plant? This question seems trivial, since we are accustomed to calling such organisms saprophytes in contrast to the "true" parasites. But the question appears in a different light when we think of the fact that according to the simple nutrition-inhibition concept such an organism should be more virulent than a pathogen which can be cultivated only on special nutritional media or even only on a living host.

To explain this situation by attributing the pathogenic inability of a saprophyte to the "natural" resistance of a living host tissue would be the same as an attempt to explain daylight by saying the sun is shining. Such a statement makes sense only when we can say what the resistance of the living host tissue is based upon. We have seen that the number of microorganisms which can penetrate into the tissue of a given host plant without producing the above described HR is negligibly small in comparison with the number of potential pathogens. This fact suggests that a pathogenic relationship can be established only when HR is not released by the infection. We should postulate such a mechanism for each normal living tissue. Noll's (1949) studies on the behavior of wheat leaves toward *Penicillium glaucum*, a common saprophyte indicate that this thought might not be too misleading—namely when the wheat leaves are exposed to a temperature of 50°C. for 30 to 50 seconds, the fungus can penetrate into the leaf tissue and there elicit a local decay. After a certain period of time, the leaf regains its resistance, i.e., the fungus stops its growth again. But, at the same time histological changes occur in the immediate vicinity of the tissue areas occupied by the fungus, which have much in common with those that are observed after a leaf is infected with a parabiotic rust race. Isn't perhaps the defense mechanism functioning here that we have postulated as the classic hypersensitivity reaction? This question has not been answered as yet with certainty. Only more experimental work can give an answer to it.

7 FUNGITOXICITY IN PLANTS

Two matters appear to be of prime importance in the physiology of fungitoxicity. The first concerns the ways that fungus cells are affected by fungicides and the second involves the factors responsible for selective action of fungitoxic compounds. The two are related and must be discussed together because the same kinds of cellular structures and physiological processes are often involved in both. It will be the object of this chapter to examine the morphological features and vital activities of fungal cells as they are related to fungitoxicity and specificity. The discussion will not be confined to plant pathogens and common fungicides but will include other organisms and other toxicants whenever they provide more appropriate illustrations of the points to be made.

Fungicidal physiology begins when a fungus cell and a toxicant are associated in the same environment. Theoretically, at least, a toxicant need not act directly on the fungus cell in order to affect its growth and development. Removal or alteration of a metabolite in the environment or inactivation of an extracellular enzyme would be sufficient in certain instances to prevent growth or to render an organism inffective as a plant pathogen. However, in the case of the fungicides presently used in plant disease control, direct action on the cell almost surely accounts for their toxicity. The fact that protective fungicides must act on spores that generally contain an internal food supply sufficient to carry them through the primary stages of infection of the host emphasizes the operation of direct rather than indirect effects. Furthermore, most fungicides developed in recent years have been discovered in screening procedures involving germination tests in which dependence on extracellular substances, except water, is at a minimum. If we accept as fact that the majority of fungicides must come in direct contact with the cells they affect and that in most instances they must enter these cells, we should first consider how fungicides, generally characterized by low solubility in water, are mobilized and how they negotiate the possible barriers that may separate them from their ultimate sites of action.

I. MOBILIZATION

Several theories have been advanced to explain how deposits of relatively insoluble toxicants on leaf surfaces or glass slides are made available in quantities sufficient to kill fungus spores. Among the agents and mechanisms that have been proposed as being involved in mobilization are atmospheric moisture, plant excretions, spore secretions, and ability of spores to accumulate toxicants from very dilute solutions.

On the subject of mobilization, which has been covered fully by Horsfall (1956), suffice it to say that Bordeaux mixture and other insoluble copper toxicants have been the subjects of the major portion of the early investigations in this area. Most of the early workers favored CO_2 and ammonia dissolved from the atmosphere in rain water and dew as the solubilizing agents for Bordeaux mixture. Curtis (1944) proposed that guttational water may play a role in increasing phytotoxicity of insoluble copper fungicides and showed that water of guttation when added to deposits on glass slides increased the fungitoxicity of Bordeaux mixture and of other copper fungicides presumably by increasing their solubility. Guttational or meteoric water at most must play only a supplemental role in mobilization of Bordeaux deposits because on glass slides in the laboratory dried deposits of Bordeaux mixture are toxic to spores applied in drops of distilled water.

On glass slides, and possibly also on foliage, secretions from spores may be important in mobilization of insoluble copper fungicides. McCallan and Wilcoxon (1936) showed that secretions from fungus spores increase the toxicity of Bordeaux deposits and that those spores most active in secretion were most sensitive to this fungicide. Malic acid as well as amino acids in the secretions from conidia of *Neurospora sitophila* are apparently responsible for an increase in toxicity. Wain and Wilkinson (1946) also found that filtrates from suspensions of spores of *N. sitophila* are active in solubilizing and increasing the toxicity of deposits of Brodeaux mixture. Horsfall et. al. (1937) and Marten and Leach (1944) showed that toxicity of copper oxides is related to their solubility in aqueous glycine solutions.

While many fungitoxicants are only very slightly soluble in water, the amount of toxicant entering water from residues without the aid of solubilizing agents is probably more than sufficient to prevent germination of sensitive spores, especially when the concentration of spores is low. With less sensitive fungi or with high concentrations of spores, however, the rate of solubilization may be a factor in limiting effectiveness. The ability of spores to accumulate large quantities of toxicants from dilute solutions is probably by far the most important factor in explaining the effectiveness of toxicants of low solubility. It has been recognized for some years, for example, that conidia of *Plasmopara viticola* and of *Monilinia fructicola* can accumulate copper to a concentration much greater than that present in the ambient solution. It is now recognized that limited solubility in water is an essential characteristic of many types of effective fungicides.

II. UPTAKE

With the advent of radioisotope tracer techniques, the problem of studying the accumulation of toxicants by fungus cells was facilitated and these techniques have been used in extensive studies of the uptake of fungitoxicants at the Boyce Thompson Institute for Plant Research. Among the toxicants studied were dichlone (2, 3-dichloro-1, 4-napthoquinone), 2-heptadecyl-2-imidazoline, and ions of silver, mercury, cadmium, zinc, and cerium. It was shown that the major portion of a toxicant ultimately taken up by a spore is removed from solution during the first 30 seconds of exposure and that affinity of conidia for these toxicants is very great. 2-Heptadecyl-2-imidazoline, for example, is accumulated in some instances to the extent that the concentration associated with the spores is 10,000 or more times greater than the original concentration in the solution in which the spores were suspended. Once these toxicants are taken up they are not readily removed by washing the spores with distilled water. The response of the spores is not dependent solely on the concentration of toxicant in solution, but on the amount of toxicant per spore, a factor which has not been fully appreciated by some investigators. Considerable quantities of these toxicants are required to inhibit germination by 50%. Compounds such as 2-heptadecyl-2-imidazoline with a low innate toxicity probably owe their usefulness mainly to their ability to accumulate in large quantities in the biophase.

Large quantities of cationic toxicants such as copper and silver are also taken up from dilute solutions by dormant ascospores of

N. tetrasperma. The evidence indicates that these toxicants are adsorbed at the surface of the dormant ascospores and they do not penetrate to susceptible sites until the spores begin to germinate. In the case of 2, 4-dichloro-6-(*o*- chloroanilino)-*s*-triazine and derivatives, Burchfield and Storrs (1957) have shown that about 1500 μg. of toxicant per gram of conidia of *N. sitophila* must be accumulated to produce 50% inhibition of germination.

The uptake of members of an homologous series of *N-n*-alkylethylenethioureas by yeast cells and by conidia of *M. fructicola* was investigated by Ross and Ludwig (1957). Toxicity in this series is related to physical rather than to chemical properties and the differential toxicity of the various members of the series is explained simply by the nature of their partitioning between the biophase and the ambient solution. As the length of the alkyl side chain is increased up to and including the octyl homologues, the proportion of toxicant taken up by the cells increases and so does toxicity. The aqueous solubility of ethylenethiourea is so high that it is not partitioned to any appreciable extent into the biophase and thus it is ineffective as a toxicant.

It is apparent that many fungicides in current use must be accumulated by fungal cells in relatively large quantities before their toxic effects are evident. On a microgram-per-gram basis, they are much less effective than penicillin, certain insecticides, and other biocidal agents.

One of the primary problems related to the uptake of toxicants is that of distinguishing between adsorption at the surface of the cell and entrance into the cell. Some aspects of this problem are discussed in a later section.

III. SURFACE OF THE CELL IN RELATION TO TOXICITY

The cell wall and possibly the outer surface of the cytoplasmic membrane are vital areas in direct contact with the external environment. These structures may represent sites of toxic action. Compounds that act at these sites do not need to negotiate differentially or semipermeable membranes and thus may differ in their physical properties from those that act within the cell. However, the principal significance of the cell wall and cytoplasmic membrane for those toxicants that act inside the cell may be that they offer resistance to penetration.

A. The Cell Wall

The cell walls of fungi are complex and variable structures. Among the substances reported to make up the cell walls of fungi are chitin, cellulose, glucan (yeast cellulose), pectin, callose, mannas, lignin, proteins, and lipids. Although all of these materials are not necessarily present in the wall of any particular fungus, lipids, protein, and one or more of the skeletal compounds, such as chitin, are generally present. Chitin is especially characteristic of the cell walls of the filamentous fungi. The materials comprising the cell walls are probably intimately associated and mutually protected from the effects of external enzymes and other agents. X-ray evidence indicates that, as a rule, the materials are linked by easily hydrolyzable bonds to form complex. Thomas (1928) found that certain components of the walls of *Fusarium* cells protect other components from the action of chemical reagents.

1. *Interference with Structure and Synthesis of the Cell Wall.* Disruption of cell wall structure, interference with the process of cell wall extension during growth or interference with synthesis of cell walls represent possible mechanisms of fungicidal action that have received relatively little attention in the past. However, the recent demonstration that penicillin can induce the formation of cell-wall-free protoplasts of *Escherichia coli* should emphasize the possible involvement of the cell wall in fungitoxic action. Lederberg (1957) proposed that the bactericidal effect of penicillin in ordinary media may be explained sufficiently on the basis of induced osmotic fragility, probably resulting from the failure of growing cells to form new walls. Results of Goksøyr's (1955) work also point toward interference with synthesis of walls as a possible basis of toxicity of dithiocarbamyl compounds to baker's yeast. Brian (1949) suggested that griseofulvin affects either directly or indirectly the cell wall or the process of extension of the wall. This antibiotic characteristically causes swelling and various types of distortion of the growing tips of fungal hyphae. He made the interesting observation that among the many species of fungi that he tested (representative of all of the orders), only those reported to have chitin walls are affected. It is entirely possible that other fungitoxicants which cause bursting of fungal cells or swelling of germ tubes and hyphal tips may do so by interference with cell wall synthesis.

There is some evidence that emergence of a germ tube from some fungal spores is preceded by a localized plasticizing of the cell wall through some metabolic process. It has been noted in this laboratory that conidia of *Fusarium roseum* in shake culture become sticky and tend to clump together just prior to germination, indicating that some alteration of the cell wall has taken place. Falcone and Nickerson (1956) have isolated a mannan-protein complex from the cell walls of *Candida albicants*. The protein in this complex contains S-S linkages that can be reduced by a "cell division" enzyme. It was suggested that localized enzymatic reduction of the protein disulfide converts a portion of the "vulcanized", cross-linked wall fabric into a form capable of plastic deformation through which a bud can emerge. A similar change may be necessary for the emergence of germ tubes from certain conidia. Toxicants that react with constituents of the walls of spores, or inhibit enzymes which alter the walls, may prohibit the emergence of germ tubes.

Synthesis of certain components of the cell walls of fungi and the processes that control the physical properties of the walls are probably distinct from those that occur in higher plants. Chitin, for instance, seems to be the main skeletal material in the cell walls of most filamentous fungi and its synthesis involves metabolism that might be selectively inhibited in fungi without injury to the higher plant. The chitin polymer is probably formed at the surface and the enzymes involved would likely be readily accessible to toxicants. Interfering with synthesis of cell walls of plant pathogens would appear to offer greater promise as a means of plant disease control than would an approach directed toward damaging the existing cell wall structure. Analogues of glucosamine might prove to be useful in the selective inhibition of plant pathogens.

2. *Permeation*. The classical point of view has been that the cell wall plays a relatively insignificant role in limiting penetration of compounds of low molecular weight and there is evidence that this may often be the case. Permeability of cell-wall-free protoplasts to low molecular weight compounds, such as sucrose, urethane, and phosphate, is reported to be essentially the same as for intact cells. Owens and Miller (1957) found insignificant quantities of 2-heptadecyl-2-imidazoline, dichlone, and Ag, Zn, Cd, Hg, and Ce ions in the cell wall fractions of spores of *N. sitophila* that had been treated with these toxicants prior to disintegration.

On the other hand, the cell wall may prevent certain substances from reaching the protoplasts. The degree of protection afforded by the walls undoubtedly varies greatly with the species of fungus. Conidia of *Erysiphe polygoni* contain large amounts of water which they do not lose readily even in a dry atmosphere. They also contain large quantities of lipids and Yarwood (1950) has suggested that fatty materials deposited in the cell walls may limit their permeability to water and toxicants such as copper sulfate. Conidia of this fungus will germinate on the surface of 10% copper sulfate solutions, although copper sulfate is toxic to the hyphae. Such substances as gelatin, peptone, inulin, an divalent succinate ions reportedly do not permeate the outer regions of cells (cell wall) of baker's yeast.

Evidence that the cell walls of higher plants may protect their protoplasts from certain fungitoxicants is provided by the work of Ross and Ludwig (1957). They found in an homologous series of *N-n*-alkylethylenethioureas that maximum phytotoxicity falls near the amyl homologue when structures such as seed coats, cell walls, and middle lamellae are present, but when these barriers were removed, the peak of phytotoxicity occurred near the octyl homologue. Uptake of these compounds by discs of potato tuber and by pure cellulose (filter paper) increased in a geometric manner as the length of the alkyl side chain increased up to and including the octyl homologue. These data indicate adsorption of the compounds by cellulose and the authors suggested that adsorption of the higher homologues by the cell walls, seed coats, and middle lamellae probably limits their phytotoxicity. Uptake by yeast cells increased in an arithmetical progression as the length of the alkyl side chain was increased and, as with the exposed protoplasts, maximum toxicity was reached at the octyl homologue. Apparently the walls of baker's yeast, composed primarily of glucan, mannas, proteins, lipids, and a small amount of chitin, do not protect their protoplasts from the higher homologues of the series.

The protection that the cell wall provides against high molecular weight substances, such as enzymes, is probably of more significance than protection that it affords from toxicants that are of comparatively low molecular weights. Bacterial protoplasts, for instance, are readily attacked by lipase, trypsin, and deoxyribonuclease, enzymes to which the intact bacterial cells are quite insensitive.

B. The Cytoplasmic Membrane

The structure of the cytoplasmic membrane, its significance as a permeation barrier, and its metabolic functions still remain obscure despite considerable research and speculation. It is generally assumed from indirect evidence that the cytoplasmic membrane is lipoprotein in nature. A generalized pattern of the membrane, consisting of a layer of lipids two molecules thick with a monomolecular layer of protein on the inner and outer surfaces, has been proposed by Davson and Danielli (1943). The cytoplasmic membrane must be considered in fungitoxicity because it may act as a permeation barrier to toxicants or the toxicants may act at this site to inhibit enzymatic activity, to block entry of metabolites into the cell, to alter permeability in such a manner as to cause the loss of cellular constituents, or to cause an irreversible physical disruption of the membrane structure.

1. *Permeation*. It is generally recognized that lipid solubility, or some property closely correlated with it, is important in determining how readily a compound permeates a cell, but that permeation is also influenced in some way by the molecular size of the permeants. Inability to permeate the cytoplasmic membrane is sometimes invoked as a possible explanation for the failure of otherwise likely compounds to function as fungicides.

Among the principal reasons for assuming that the cytoplasmic membrane is involved in regulating the entrance of toxicants into the cells are the correlations between toxicity and the proper degree of lipid solubility of compounds and the fact that neutral molecules of many toxicans are frequently more effective than the corresponding ionized forms.

In toxicants exerting their effect through chemical mechanisms, i.e., structurally specific compounds possessing one or more active chemical groups that combine with specific cell constituents, the proper degree of lipid solubility is thought to be a physical property required for permeation to the susceptible sites within the cell and is not involved in their ultimate activity. Thus, when increased toxicity of this type of compound is achieved by the addition of inert lipophilic substituents to the molecule, it apparently results from more ready permeation. This is reported to be the case with the nitrosopyrazoles, in which the nitroso is regarded as the active group. Addition of inert hydrophobic substituents to the pyrazole ring increased the oilwater partition coefficient as well as toxicity, and the increased toxicity in this case was attributed solely to the increased ability of the compounds to permeate the cell membrane. This appears to be the case with a number of organic fungicides. With the 8 quinolinols, for example, lipid solubility as well as ability to chelate is required for activity. It has been postulated that the—$SCCl_3$ group in the captan molecule functions in permeation, but there is now evidence that this group may be the toxiphore.

A somewhat different relationship betwen permeation and lipid solubility seems to characterize the chemically inert, structurally non-specific toxicants that act by physical, rather than by chemical, mechanisms. These are compounds that do not owe their toxicity to a specific chemical group but rather to their ability to accumulate in the biophase. The *N-n*-alkylethylenethioureas are representative of this kind of toxicant. The relative effectiveness of such substances is related to their partition coefficients between fat-like substances and water. Structurally nonspecific compounds of diverse types have approximately the same potency when they are present in the same proportional saturation in the medium in which they are applied even though their molar concentrations in which they are applied even though their molar concentrations in that medium may differ widely. This is known as Ferguson's Principle and it seems to apply to all compounds which owe their activity to certain physical-chemical properties that favor their accumulation in the biophase. An increase in activity of such compounds as the oil-water partition coefficient increases does not necessarily mean that their ability to permeate membranes was the limiting factor in toxicity. Johnson et. al. (1954) point out that the same type of bond is formed when such substances dissolve in oil as when they combine with hydrophobic groups of enzymes or other proteins. The correlation between lipid solubility and potency also applies to purified enzymes *in vitro* in the absence of lipids and cellular structure which suggests that hydrophobic portions of proteins may be the sites of action. These sites may occur at the cell surface. However, in the intact cell, lipid barriers at the membrane may have to be negotiated before susceptible proteins are encountered and thus partition between lipid and water may be a measure of their ability to penetrate membranes as well as a measure of their ability to combine with hydrophobic groups in protein molecules. Since water-lipid solubility, surface activity, and absorbability are somewhat parallel phenomena, one wonders whether ease of penetration of the cell membrane is the only significance of the proper degree of lipid solubility even in the case of structurally specific toxicants.

The biological activity of the ions of weak acids and bases is frequently lower than that of the corresponding undissociated molecules. The differences in activity usually have been attributed to difficulties in permeation by the ionic forms. Permeation is affected by the greater relative size of ions, due to hydration, by their charge resulting either in repulsion or fixation by adsorption, and to their generally lower solubility in lipids. Simon and Beevers (1952) report that pH changes in the range 3 to 6 have no effect on the concentration of iodoacetic acid (pK = 3.1) required to produce 50% inhibition in the fermentation of glucose by zymase, but the concentration required to halve the fermentation rate of intact yeast cells was thirteen times greater at pH 6.0, where the compound existed almost entirely in the ionic form, than at pH 3.0 where the concentrations of ions and neutral molecules were approximately equal. This evidence suggests that the cytoplasmic membrane restricts permeation of the ionic form of this toxicant. On the other hand, it is the ionic forms rather than the neutral molecules that are responsible for toxicity of some compounds. This is the case for the acridine antibacterials. It is probable, however, that they act at the surface of the cells rather than internally, because the addition of aliphatic side chains to the molecule to aid permeation reduces their antibacterial action.

It might be pointed out here that mitochondrial and nuclear membranes may also constitute permeation barriers that may be of considerable significance in fungitoxicity. Both mitochondria and microsomes are rich in lipids and there is evidence that the membranes of mitochondria are semipermeable.

There is considerable variance of opinion on the significance of the cytoplasmic membrane as a permeation barrier to normal metabolites and to toxicants. Mitchell and Moyle (1956) hold that the cytoplasmic membrane of *E. coli* is relatively impermeable even to phosphates and sodium chloride. Conway and Downey (1950) concluded that arabinose, galactose, ions of Na, Cl, and K, and many other substances enter the outer region of baker's yeast, but are apparently barred by the cytoplasmic membrane or enter only very slowly into the cell. Mandels (1953) achieved selective destruction of surface-located carbohydrases in spores of *Aspergillus luchuensis* and *Myrothecium verrucaria* with 0.1 *N.* HCl without significantly affecting other metabolic processes or the subsequent rate of growth. It is difficult to conceive of such selective action of hydrogen ions without the participation of a membrane that excludes most of them from the cytoplasm. Conidia of *Erysiphe polygoni* germinate well on a 10% copper sulfate solution and one must conclude as Yarwood (1950) did, that this resistance is best explained as a failure of the toxicant to permeate into the spores. Lipids in the cells wall, however, may be more important than the cytoplasmic membrane in excluding the toxicant in this case.

On the other hand, there is evidence that the cytoplasmic membrane in certain fungi and other organisms may not constitute a very significant barrier to the permeation of metabolites and toxicants of low molecular weight. Some plant physiologists have experimental evidence to indicate that a cytoplasmic membrane may not impede the permeation of ions into the cytoplasm of cells of roots of higher plants. Similarly, Roberts et. al. (1955) concluded from their experiments with *Escherichia coli* that permeation of many compounds of low molecular weight into most or all of the water space of the cell proceeds unimpeded by a membrane at the surface. Inorganic ions, phosphorylated sugars, amino acids, and glutathione are reported to permeate readily. Less extensive experiments with the yeast, *Torulopsis utilis*, indicated that there is no lack of permeability in this organism. It is well known that intact cells of many organisms do not metabolize the phosphate esters of sugars, such as fructose-1, 6-diphosphate, when they are supplied externally. However, these compounds can be isolated from living cells and they are readily metabolized by cell homogates. It would appear that in most cases the failure of intact cells to metabolize phosphorylated sugars must be one of permeability, but, according to Roberts, et. al. (1955), some other reasons must explain the failure of *E. coli* to metabolize fructose-1, 6-diphosphate.

Although the experiments of Miller et. al. (1954), concerning the uptake of toxicants by fungus spores, were not designed specifically to show whether the toxicants actually enter the cytoplasm, ready permeation was inferred. In this connection, McCallan (1957) indicated that permeation of the fungus spore may not be the obstacle it was once considered to be. In further studies of uptake, Owens and Miller (1957) attempted to determine whether the toxicants, dichlone, glyodin base (2-heptadecyl-2-imidazoline), and ions of five heavy metals, actually enter the cytoplasm of conidia of *N. sitophila* and of *A. niger*. When conidia were exposed to sublethal doses of the fungitoxicants and then disintegrated by sonic treatment, essentially all of the toxicant that was taken up from the medium was found in the supernatant fraction and in the fraction containing particulates corresponding in size to the mitochondria and microsomes. Even though the spore walls of *Neurospora* account for 35% of the weight of the conidia, the fraction containing cell wall fragments retained only insignificant amounts of any of the toxicants.

Experiments with uptake and permeation have dealt largely with successful toxicants and fungi sensitive to them and they indicate that the toxicants are capable of entering and exerting their effect in the cytoplasm of cells. Experiments of this type designed specifically to compare uptake of a toxicant by sensitive or immune organisms, should indicate to what extent differential uptake and permeation are responsible for selective toxicity. This is an area of investigation that has as yet received little consideration.

One might conceive of the following differences between resistant and sensitive species or between pathogen and host that would account for their differential uptake and response to a toxicant: (1) the toxicant does not permeate into the cytoplasm of either the resistant or susceptible species, but acts at sites on the surface of sensitive cells. Resistant species do not possess such sensitive sites at this locus; (2) the toxicant permeates readily into both resistant and susceptible cells, but only the latter possess highly sensitive sites internally; and (3) both resistant and susceptible species possess equally susceptible sites internally, but differentiation permeability to the toxicant at the cytoplasmic membrane or at other levels in the cell accounts for selective toxicity. It should be recognized, of course, that within these categories the differences need not be absolute, but may be relative and that various combinations may occur.

2. *Interference with Activities at the Cytoplasmic Surface*. Species may differ more widely in their surface structure and activity than they do internally. Metabolic activities occur at the surfaces of cells. Digestion of extracellular, nonutilizable substrates into forms that may be absorbed by the cell, active transport, self-maintenance, and self-replication are biochemical functions attributable to the cell surface. Among the enzymes reported to be located at the surface of fungus cells are certain carbohydrases, phosphatases, and ascorbic acid oxidase. It has been suggested that the protein of the cell membrane functions not only by endowing the membrane with certain structural properties, but that it also has enzymatic properties. Both extracellular enzymes and cell wall material must be synthesized at the surface of the cell or else some mechanism must be assumed that will allow passage of these substances of high molecular weight through the plasma membrane.

It is not known how frequently toxicity is a result of interference with vital activity at the cell surface, because it is usually difficult to distinguish between interference this site and that occurring internally. There is evidence, however, that some toxicants may act primarily by interfering with activity at this locus. For instance, the predominant, acute, toxic action of uranium to yeast is reported to involve combination with multiphosphate compounds at the cell surface. The uptake of fermentable sugars is blocked, but metabolism of other substrates remains essentially unaffected. Endogenous respiration and oxidation of acetate, pyruvate, lactate, and alcohol are insensitive to a concentration of this inhibitor that blocks sugar uptake.

In *N. sitophila*, uranium has an effect similar to that noted in yeast. Concentrations of uranyl nitrite that completely inhibit respiration of glucose do not affect the oxidation of acetate, ethanol, or glycerol. The conclusion was drawn that uranium must block reactions near the cell surface which are essential for glucose transport.

Goksoyr (1955) concluded that certain dialkyl dithiocarbymyl compounds inhibit the growth of *Saccharomyces cerevisiae* by interfering with thiol-activated enzymes located at the cell surface. His data show that the only decrease in metabolic activity (when related to degree of growth) that is brought about by these dithiocarbamates is to be found in processes leading to synthesis of cell wall material which, according to Rothestein (1954), most probably occurs at the surface of the cell which is in direct contact with the cell wall.

Selenate competes with sulfate for absorption sites on the cell membrane of algae. Selenate can be largely excluded from algal cells if sufficiently high concentrations of sulfate are present. Growth of histidine-requiring mutants of *N. crassa* is inhibited when certain combinations of amino acids are provided together with histidine. Growth of the wild type of this species is not affected by such treatment, nor is growth of the histidine-requiring mutants if they are allowed to take up histidine prior to the addition of the inhibiting amino acid combinations. It was shown that this inhibitory combination of amino acids prevents uptake of histidine by both the wild type and the histidine mutants. The former are able to synthesize the histidine they need while the latter are not. This seems to be an excellent illustration of toxicity based on interference with the entrance of a metabolite into the metabolic system. A similar mechanism of interference may explain why certain toxicants that are analogues of metabolites of both host and pathogen are toxic only to the pathogen. Woolley (1952) discussed instances wherein an antimetabolite is toxic only when the organism requires the corresponding metabolite from an external source. Since many fungi are apparently capable of invading the host plant in the absence of exogenous nutrients, spraying plants with compounds that interfere with uptake of metabolites would not appear to offer promise as a means of plant protection. On the other hand, cells of higher plants are able to synthesize organic metabolites that many pathogenic fungi must obtain from their environment. It is apparent, therefore, that toxicants which block the entrance of such metabolites into fungal cells might be useful chemotherapeutants.

3. *Effects on Membrane Permeability and Structure*. Some toxicants may act by disrupting the cytoplasmic membrane or by altering its permeability so that cellular constituents are lost into the surrounding medium. At lethal concentrations, silver causes conidia to lose considerable quantities of phosphorus to the ambient solution. Interference with permeability represents an important characteristic of the action of silver. Tyrocidine causes bacterial cells to leak their soluble constituents, presumably by alteration of the cellular membrane. A number of cationic and anionic surface-active agents produce the same effect as tyrocidine and their toxicity is attributed to an irreversible damage to the cytoplasmic membrane. According to Hotchkiss, ionic surface-active agents combine with oppositely charged ions at the surface of the cell and if the hydrophobic portion of the toxicant also has proper affinity for the surface, irreversible damage to the cell membrane results at very low concentrations. Soluble constituents are released from the cells and this is followed by autolysis with further release of cellular constituents. It would not be surprising if both glyodin and *n*-dodecylguanidine acetate (Cyprex) were reported to cause leakage of metabolites from fungal cells, because both are cationic, surface-active agents. Newton (1953) and Few and Schulman (1953) showed that polymyxin induces loss of cellular constituents from bacterial cells and suggested that it combines with components of the cell surface, thus altering permeability of the cells. Algal cells lose cellular constituents when treated with polymyxin. Lowry and Sussman (1956) showed that polymyxin also causes loss of cellular constituents from ascospores of *N. tetrasperma* after they begin to germinate and are susceptible to osmotic damage. It is not absolutely certain, however, that changes in permeability may not be a secondary effect of intracellular origin. Polymyxin also prevents uptake of methylene blue, presumably by blocking sites at the surface of the ascospores.

The significance of loss of soluble cellular components in fungicidal action is not entirely clear, because spores of some fungi can apparently lose considerable quantities of certain constituents when suspended in distilled water and they may lose additional quantities when treated with sublethal doses of certain fungitoxicants without any impairment of their ability to germinate. Spores of *N. sitophila*, for example, when treated with an *s*-triazine lose cellular constituents such as phosphorus, and this loss is attributed to changes in permeability induced by the toxicants, but the spores germinate well and it appears unlikely that effects on premeability alone are primarily responsible for toxicity of these fungicides.

IV. DAMAGE OF PHYSICAL STRUCTURE AND ORGANIZATION

In addition to plasma membranes there are other structures and a great deal of organization that must be maintained in a living cell. Irreversible disruption of structure or organization results in malfunctioning and death of cells. Kauzmann (1954) presents a discussion of the types of weak, secondary, intramolecular bonds responsible for maintaining the structural patterns of enzyme proteins. These include hydrogen bonds, ionic bonds, van der Walls' forces, cystine disulfide bonds, and "hydrophobic bonds." The last type involves the adherence of hydrophobic amino acids residues of proteins to each other as a result of their tendency to avoid the aqueous phase. Hydrophobic bonds are considered to be particularly important in maintaining structural stability of protein molecules. Rupturing of the bonds responsible for maintaining protein structure leads to denaturation. Undoubtedly, linkages of these several types are also of major significance in joining protein molecules in the cytoplasm, in linking proteins and lipids in the cell membrane, and in linking nucleic acids with proteins in the chromosomes, mitochondria, and microsomes. Changes in viscosity of protoplasm are probably associated with the formation and breaking of these linkages. Toxicants that rupture the linkages responsible for maintaining stability of cell structure may cause irreversible physical damage and loss of organization in the cell.

Since the toxicants which act by physical mechanism have much greater affinity for lipid-dissolving materials than for water, they probably collect at hydrophobic junctions and weaken or disrupt these linkages, causing damage to physical structures. It has been suggested that organic solvents and detergents denature proteins in this manner. Dodecylsulfate is commonly used to dissociate nucleic acid from protein in viruses. Its action is probably that of disrupting ionic and other linkages joining the proteins and nucleic acids. 2-Heptadecyl-2-imidazoline is known to associate with particulate fractions of the cytoplasm of fungal cells and, at toxic concentrations, it may disrupt physical structure. On the other hand, these ionic surface-active toxicants may prevent attachment of enzymes to coenzymes or to substrates at concentrations lower than those necessary to disrupt structural features of the cell.

V. INTERFERENCE WITH INTRACELLULAR METABOLISM

While there is evidence that some fungitoxicants may act at the surface of the cell it would seem to be a safe assumption that most toxicants exert their effects internally. The vast number of enzymatic reactions involved in release of energy from substrates and in syntheses of materials in the cytoplasm undoubtly account for the major portion of vital cellular activity with which a toxicant might interfere in a lethal manner.

It is becoming increasingly clear that there is much similarity between the vital intermediate metabolism of fungi and that of other organisms. The vitamins, coenzymes, amino acids, purines, and pyrimidine, as well as the majority of other known metabolites of a vital nature, are identical in fungi and in other organisms. The fatty acids—palmitic, steraric, oleic, and linoleic—which are found in largest quantities in lipids of fungi are also abundant in lipids of higher plants. Likewise, the general metabolic pathways of fungi seem to be very similar to those of other organisms. The operation of the Krebs cycle in the filamentous fungi and in yeasts has been demonstrated several investigators. Pentose cycle reactions have been demonstrated in *Penicillium chrysogenum* by Sih, et. al. (1957) and cytochrome pigments were found in all of 47 species of fungi investigated by Boulter and Derbyshire (1957). It would seem logical that protein and nucleic acid synthesis would proceed in fungal cells in essentially the same manner as it does in cells of other organisms.

One meets with considerable difficulty in explaining selective toxicity of fungicides on a basis of differences in metabolism of host and pathogen since their general metabolic processes are apparently very similar. Unique vital processes are more characteristic of the higher forms that they are of fungi. Thus, the possibilities of selectively controlling a group of organisms by interfering with activities unique to that group are more promising with the more complex forms of life. However, in different ogranisms there are undoubtedly many minor variations in metabolic pathways and in enzymes that have the same function. These differences may be of considerable significance in selective toxicity. Lysine biosynthesis, for instance, appears to proceed by a pathway in yeast and in *Neurospora* which is different from that in *E. coli*. ε-Amino adipic acid is a precursor for lysine in these fungi, while diamino pimelic acid is a precursor in *E. coli*. Alcohol dehydrogenase and glyceraldehyde-3-phosphate dehydrogenase from yeast differ in certain of their properties from the corresponding enzymes from animal sources. The protein moiety of cytochrome *c* from heart muscle differs from that of cytochrome *c* from *Ustilago sphaerogena*. Saz et. al. (1956) proposed that an altered enzyme may explain the resistance of a strain of *E. coli* to chlortetracycline. In cell-free preparations from a sensitive strain of the organism, the organic nitroreductase was inhibited by bacteriostatic concentrations of the antibiotic while in similar preparations from resistant cells the enzyme was likewise resistant to the antibiotic. One of the differences noted was that the enzyme from the resistant strain contained firmly bound, conjugated flavin, while the flavin of the nitroreductase from the sensitive strain was easily dissociable.

Differences in metabolic pathways and enzymes may be of considerable significance in determining the effectiveness of toxicants that are highly specific in their modes of action. They can, however, hardly account for the selective control of pathogenic fungi that has been attained with toxicants that are not selective in their modes of action, but are capable of reacting with common chemical groups, —SH for example, that may occur at many points in most living cells. In these cases differences in permeability and in a number of other factors are probably of more significance than differences in susceptibility of enzymes. There is little doubt that the heavy metal fungitoxicants and possibly some of the organic fungicides used in the practical control of plant diseases would be about equally destructive to the metabolism of higher plants as of fungi if the concentrations attained in the cytoplasm of both types of cells were equivalent.

Fungitoxicants that affect cells by interference with intracellular metabolism might be considered as falling into two general groups: (A) Those analogues of metabolites that disrupt normal processes by competing with metabolites and, (B) those that affect metabolism by virtue of their ability to interfere with enzymes, coenzymes, metal activators, or substrates in some noncompetitive manner.

A. Effects of Antimetabolites

It appears that most of the toxicants that are highly specific in their modes of action are analogues of metabolites. These toxicants, because of their close similarity in structure or characteristic chemical reactivity, to normal metabolites interfere in a reversible manner with the utilization of normal metabolites by substituting for them in metabolic processes. They are often referred to as antimetabolites. Many analogues of metabolites have been reported since Woods (1940) first showed that sulfanilamide in an analogue of the essential metabolite, *p*-aminobenzoic acid.

The primary effect of such toxicants may be limited to a single reaction in some cases, although if the affected reaction is in some primary stage of metabolism, many other reactions will be affected secondarily. For instance, prevention of incorporation of any one amino acid into protein by an effective amino acid analogue prevents utilization of virtually all other amino acids as well. On the other hand, much of the primary metabolism of a cell may remain unaffected when an analogue interferes with some more advanced stage of metabolism. Sulfanilamide is an antimetabolite for *p*-aminobenzoic acid in fungi as well as in bacteria, but it does not affect oxidation of sucrose by cells of *N. crassa* or of glucose by conidia *F. roseum* at concentrations that inhibit growth. There also appears to be no affect on the processes of assimilation in resting cells of *F. roseum*, because the treated cells increase in weight to the same extent as do the untreated cells.

In some instances an analogue may participate in a curious course of events in which it is utilized by a cell in the synthesis of another analogue which is actually responsible for toxicity. Fluoroacetic acid is a case in point. It is an analogue of acetic acid and is reported to be synthesized by the cell into fluorocitric acid which interferes with aconitase and prevents the conversion of citrate to isocitrate. It may well be that the fluoroacetate is also incorporated into lipids and other cellular components that are normally built up from acetate. Selenate an analogue of sulfate, interferes with sulfate utilization by cells of *Chlorella vulgaris* and is also synthesized into selenomethionine, an antimetabolite of methionine. An analogue of tryptophan, 7-azatryptophan, does not prevent protein synthesis in *E. coli*, but does prevent the development of a number of enzyme activities. The proteins that are synthesized in this case are probably inactive analogues of the normal enzymes. These are examples of lethal synthesis, a subject that is discussed in more detail by Markham (1958).

Most of the antimetabolites that have been described, either coincidentally resemble, or have been patterned deliberately after vital metabolites that are common to many species. They are not specific in the sense that they resemble a metabolite peculiar to a particular group of organisms. Horsfall (1956) has complied a list of compounds that function as antimetabolites in fungi. Practically all of the compounds that the listed, however, are potential antimetabolites for other organisms as well and many of the compounds listed by Woolley (1952) are analogues of metabolites common to numerous species. With our present understanding of comparative biochemistry the changes of deliberately designing and synthesizing an analogue specific for a metabolite peculiar to fungi are not as good as are the chances of designing one specific for higher organisms with more complex and highly specialized activities. However, as our understanding of the details of differences between the metabolism of cells of various organisms increases, the chances of selecting or synthesizing an antimetabolite for a particular process should also increase and offer promise of highly selective control of plant pathogenic organisms.

Remarkably enough, selective inhibition of certain species of organisms without affecting others has often been accomplished with compounds that are apparently analogues of metabolites common to both the sensitive and resistant species. Woolley (1952) has provided an excellent example to illustrate this point. Pyrithiamine is an analogue of thiamine and is known to be toxic to certain higher animals, invertebrates, fungi, and bacteria, but there are many species of fungi, bacteria, and higher plants that are unaffected by this compound. There is little doubt that thiamine is a metabolite in both resistant and susceptible species. Resistant and susceptible forms are in some cases closely related and genetic differences between many of the resistant and susceptible organism must be very slight. Therefore, selective toxicity of pyrithiamine, based on gross and fundamental morphological and physiological differences, does not seem likely, especially since widely separated species have a similar degree of susceptibility. Woolley (1952) gives a thorough discussion of a number of factors that may be involved in selective toxicity in such cases.

Very few of the well-known fungicides appear to be antimetabolites and even those that have been reported to function in this fashion seem capable also of other, less specific modes of action. among those for which antimetabolitic activity has been suggested are 2-heptadecyl-2-imidazoline, dichlone, sulfur, and captan. Both guanine and xanthine antagonize the toxicity of 2-heptadecyl-2-imidazoline to *Sclerotinia fructicola* and the antagonism is of the competitive type. This would indicate that this toxicant is an antimetabolite for these purines, but it is a cationic surface-active agent and, thus, one might expect it to be capable also of other less specific effects. Germicidal activity is generally characteristic of ionic surface-active agents some of which are known to cause osmotic damage and to disrupt macromolecular structure. McCallan et. al. (1954) showed that concentrations of 2-heptadecyl-2-imidazoline that reduce oxygen consumption of certain fungal conidia by 50% also reduce their germination by about the same amount. This implies that the fungicide may affect processes other than xanthine and guanine metabolism. Dichlone resembles vitamin K in structure and Wooley (1945) demonstrated that vitamin K will reverse the fungitoxicity of dichlone in a competitive manner, but only over a very limited range of concentrations of the toxicant. It was recognized, however, that the reactive chlorine atoms in the dichlone molecule might well confer on it toxicity that is unrelated to any effect that it may exert as an antimetabolite. McCallan (1957) considered the hypothesis that sulfur may substitute for oxygen in the respiratory systems of fungi and that it might exert its toxic effects by competition with oxygen. Since sulfur actually stimulates oxygen consumption in fungi and since cyanide does not inhibit the capacity of fungi to convert sulfur into H_2S, it was considered unlikely that sulfur acts by replacing oxygen in the respiratory process. There remains the possibility, however, that sulfur may complete with other hydrogen acceptors in cells and thus distrust their metabolism, as has been suggested by Miller et. al. (1953). Hochstein and Cox (1956) demonstrated competition between captan and cocarboxylase in preparations of dried brewers' yeast. It was also shown that in conidia of *F. roseum* that were treated with captan, pyruvic acid accumulated to a greater extent than in untreated conidia as would be expected if decarboxylation of pyruvate were inhibited. Although thiamine pyrophosphate offset the effect of captan in yeast preparations, it did not protect living cells and impermeability of the cells to this coenzyme was considered as a possible explanation of its failure to protect. It was suggested that captan may owe its toxicity to interference with enzymes that require thiamine pyrophosphate.

Xanthine and guanine, oxygen, vitamin K, and thiamine pyrophosphate are metabolites that are by no means peculiar to fungi. If the toxicants, dichlone, sulfur, and captan do behave as antimetabolites in fungi, then there must be differences other than metabolic ones that are at least partially responsible for rendering these compounds more toxic to fungi than to higher plants and animals.

Actually, inhibition of metabolism by competition may be more common than is presently suspected since some toxicants not now regarded as antimetabolites may permeate so readily and have such high affinity for the susceptible enzyme that they would accumulate in cells to such an extent that the range of concentration of normal metabolite required to demonstrate the competitive nature of the inhibition cannot be attained in the intact cell. This would be the case especially when the cell is impermeable to the normal metabolite, i.e., coenzyme supplied externally.

Many cationic and anionic toxicants may act in the cell as inhibitory analogues of essential ions. This would constitute a somewhat more general type of competitive inhibition than that which is commonly associated with antimetabolites. According to MacLeod and Snell (1950), certain metallic ions which inhibit growth of bacteria might be regarded as structural analogues of other metallic ions which are essential for certain metabolic processes involved in growth.

B. Effects of Other Types of Toxicants

The vast majority of the fungicides that are used in the practical control of plant diseases appear to be capable of much more generalized effects on metabolism than are the analogues of metabolites. Results of research in recent years indicate that many of them are potentially capable of reacting with functional chemical groups of enzymes and coenzymes, with metal activators, or other components of metabolic systems, many of which are common to various aspects of metabolism. This being the case, it does not seem practical to separate metabolism into various categories such as energy release, protein synthesis, etc., and discuss how fungicides affect each of them. It does seem to the point, however, to discuss the manner in which several representative types of fungicides are reported to affect cell metabolism and enzymes. It is also appropriate to mention here some of the possible explanations of how these fungicides may be used to control pathogens without injuring the higher plants to which they are applied.

The activity of quinones against diverse types of enzymes suggests that they are capable of blocking cellular metabolism at a number of stages. Quinones are effective inhibitors of pancreatic amylase, an aminodependent enzyme, and of malt amylase, a sulfhydryl-dependent enzyme and they are also moderately inhibitory to catalase. A number of the quinones are potent inhibitors of carboxylase. In fact, their effectiveness as inhibitors of carboxylase was sufficiently correlated with their toxicity to spores of *Monilinia fructicola* that Foote et. al. (1949) were led to suggest that *in vitro* carboxylase inhibition might be used as a preliminary screen for selecting quinones for use as fungicides. A number of other enzymes, including urease, pancreatic lipase, proteinase, and phosphatase, are also inhibited by certain quinones.

Dichlone, which is 2, 3-dichloro-1, 4-napthoquinone, reacts with coenzyme A *in vitro* and it was suggested that its extensive effects in disrupting cellular metabolism in *N. sitophila* may result from its blocking of many reactions in cells that require coenzyme A. The concentrations of chloranil and of dichlone that are required to inhibit 50% of the oxygen consumption by conidia of some species of fungi are essentially equivalent to the amounts required to prevent 50% of their germination, but other fungi require greater quantities to inhibit oxygen uptake by 50% than are required to give an equivalent inhibition of germination. This suggests that retardation of growth in the two cases may result from effects on different phases of metabolism. Some degree of selectivity for a particular phase of metabolism within sensitive cells might be expected as a result of variations in enzyme sensitivity or of the distribution of the toxicant within the cell, but it seems most likely that a number of metabolic processes would be affected with almost equal facility by growth-inhibiting concentrations of quinones. Since the quinones are capable of combining with both —SH and amino groups, of oxidizing —SH groups, and of changing redox potentials, their selectivity toward various organisms, as Horsfall (1956) has concluded, probably rests mainly on two factors—permeation and detoxification. Evidence presented by Rich and Horsfall (1954) indicates that polyphenol oxidases in fungi may be significant in detoxifying phenols and quinones.

At concentrations sufficient to prevent germination and growth, captan, *N*-(trichloromethylthio)-4-cyclohexene-1, 2-dicarboximide, virtually eliminates uptake of oxygen by conidia of *M. fructicola* and it is as effective in reducing consumption of oxygen by spores of *Alternaria solani* as it is in suppressing their germination. Oxygen uptake by dense suspensions of conidia of *F. roseum* is severely reduced by captan at a concentration of 10^{-4} moles per liter. Hochstein and Cox (1956) showed that pyruvate accumulates in cells of this fungus when they are exposed to captan and that yeast carboxylase activity *in vitro* is inhibited. Thus, it appears that effects on early stages of metabolism (assimilation of substrate and carbohydrate oxidation) may be sufficient to account for toxicity of captan to some fungal cells. Its effects are not limited to fungi. Captan is toxic to cells of higher plants if it gains access to them in sufficient concentration. It is quite toxic to bean and tomato plants even at low concentration (1–5 p.p.m.) when it is applied to the roots in liquid culture. Apparently it is capable of affecting metabolic systems in some higher plants similar to those that it affects in fungi because Dugger et. al. (1959) have shown that it causes accumulation of pyruvate in pea root tips and that it strongly inhibits pyruvic oxidase and α-ketoglutaric oxidase activity in preparations of mitochondria from lupine plants. They also reported that captan inhibits hexokinase from yeast and the oxidation of ribose-5-phosphate by extracts from peas. Lukens ad Sisler (1958) have presented evidence to support the contention that captan owes its ultimate toxicity to the reactive —$SCCl_3$ group that is released or to thiosposgene that is formed following cleavage of the captan molecule by sulfhydryl groups in commonly occurring cellular components. One would not expect a high degree of specificity from these reactive derivatives. It would appear, therefore, that such specificity as captan possesses must be attributed to differences in its ability to penetrate and accumulate in cells and possibly also to deferences in the facility with which cells of different species are able to release the —$SCCl_3$ group from the imide portion of the molecule.

An understanding of the way in which such dialkyldithiocarbamate fungicides as thiram (tetramethylthiuram disulfide), ziram (zinc dimethyldithiocarbamate), and their relatives affect cellular metabolism has been complicated by the tendency of these fungicides to be less toxic when used at certain intermediate concentrations than when used at either lower or higher concentrations. Dimond et. al. (1941) first reported this kind of behavior for thiram and it has since been observed by a number of investigators with thiram and related compounds. The range of intermediate concentrations at which growth of fungi occurs has been designated by Sijpesteijn and van der Kerk (1956) as the zone of inversion growth. Occurrence of this phenomenon appears to be regulated by the concentration of certain heavy metals in the medium or in association with the cells. With *Saccharomyces pastorianus* as the test organism, the zone of inversion growth characteristic of sodium dimethyldithiocarbamate was almost completely abolished by addition of certain trace metals in sufficient quantity to the medium in which the yeast was grown. Zinc appears to be the most critical of the metals in this case. Areas of metabolism affected by concentrations of fungicide on the lower side of the zone of inversion growth are probably different from those affected by toxic doses on the upper side of the zone of inversion growth.

α-Ketoglutaric acid accumulates in the mycelium of *F. roseum* when it is treated with ziram and pyruvic acid accumulates in cultures of *A. niger* and *Penicillium italicum* when certain concentrations of sodium dimethyldithiocarbamate are present. α-Keto acids antagonize the toxicity of this fungicide toward *A. niger*, but they do not offset its toxicity to certain other fungi. It was thought originally that the pyruvic acid that accumulated in the presence of the fungicide might play an important role in initiating the phenomenon of inversion growth, but it now appears that this is not the case.

Accumulation of pyruvate and α-Ketoglutarate, which are intermediates of carbohydrate metabolism, suggests that dithiocarbamates affect respiration, and such an effect has been observed. Thiram suppressed oxygen uptake, carbon dioxide production, and germination of conidia of *F. roseum*, and of *Myrothecium verrucaria*. Thiram's close relative, antabuse (tetramethylthiuram disulfide), prevents growth of *Staphylococcus aureus* at quite low concentrations and inhibition of growth is directly correlated with suppression of respiration. At concentrations that prevent growth of *F. roseum*, ziram suppresses respiration by about 50%, but higher concentrations have no further effect on respiration. McCallan et. al. (1954) showed that respiration of condidia of *Monilinia fructicola* is more sensitive to ferbam than is their germination, but with spores of some other species of fungi the opposite is true. At a concentration that prevents growth of *Saccharomyces cerevisiae*, sodium dimethyl dithiocarbamate was demonstrated by Gokøyr (1955) to depress oxidation of glucose by about 20%, but oxidation of acetate is somewhat more inhibited than is that of glucose.

These examples serve to demonstrate that at concentrations which are toxic to fungi, dithiocarbamates often affect respiration, but that the extent of the effect and the relationship between suppression of respiration and suppression of germination of spores and growth of fungi varies with the particular fungicide and with the species of fungus that serves as the test organism. There is not now sufficient evidence in the case of any of the dithiocarbamates to indicate conclusively that inhibition of growth of fungi is the direct result of suppression of respiration. In those cases where growth-inhibiting concentrations have little or no effect on respiration it would seem to be a safe assumption that the primary basis of toxicity is to be sought elsewhere.

When it was discovered that the toxic effect of some of the dialkyldithiocarbamates is overcome in part by addition of histidine or of certain other imidazole derivatives, it was suggested that they may owe their toxicity to interference with synthesis of histidine or with some vital process in which it is required. However, since cysteine is also somewhat antagonistic and, like histidine, is a chelator, it now appears that the protection that they afford is most probably related to their ability to complete for heavy metals involved in toxicity of the dithiocarbamates.

There is a considerable literature on the effects of dialkyldithiocarbamates on enzymes from various sources. Among the enzymes reported to be affected are: copper-containing enzymes such as catechol oxidase, laccase, and ascorbic acid dehydrogenase the sulfhydryl enzymes—aldehyde dehydrogenase, triosephosphate dehydrogenase, and succinoxidase. In addition, Owens (1953) demonstrated that pancreatic amylase, malt amylase, polyphenoloxidase, and catalase, which respectively represent amino-, sulfhydry-, copper-, and iron-dependent enzymes are all inhibited by dithiocarbamic acid derivatives. Thus, it is apparent that these fungicides possess the potential of interfering with a number of metabolic processes. However, the fact that they inhibit a wide variety of enzymes when isolated in the laboratory does not necessarily imply that they owe their toxic action to effects on these enzymes in the intact cell. This is illustrated by Gokøyr's (1955) observation that certain dithiocarbakyl fungicides are strongly inhibitory to yeast succinoxidase *in vitro*, but have little or not effect on oxidation of succinate by yeast cells.

Goksøyr's (1955) extensive studies of the effects of dithiocarbamyl compounds on *S. cerevisiae* indicated that in intact cells the major areas of metabolism, such as respiration, protein synthesis, and synthesis of nucleic acids, are not primarily affected by the dithiocarbamates. He concluded that these fungicides most probably act at the surface of cells, affecting in varying degrees a rather large and diverse group of enzymes at that location. It remains to be seen whether this is the case for fungi in general, in view of the reports of the accumulation of keto acids in cells treated with these toxicants. Goksøyr finally concluded that anatomical differences between cells of different organisms must play a more important role than do physiological differences in determining selectivity of the dialkyldithiocarbamyl fungicides.

The bisdithiocarbamates, the best known fungicidal representatives of which are nabam, zineb, and maneb, apparently operate by mechanisms that differ from those of the dialkyldithiocarbamates. Presently available evidence indicates that they are converted to isothiocyanates that inactivate essential —SH systems of cells.

Turning our attention to a relatively new group of fungicides, the -*s*-triazine derivatives, we find that they possess the potential of arresting a variety of vital biochemical reactions in cells. It was shown that 2, 4-dichloro-6-(*o*-chloroanilino)-*s*-triazine, for instance, will combine with amino acids, glutathione, para-amino-benzoic acid, proteins, pyridoxine, pyridoxamine, and niacin. In view of these capabilities, it was concluded that any selectivity that these compounds may demonstrate with regard to sites of action in cells is probably dependent upon relative reaction rates rather than upon absolute specificity for a single receptor site. Since they are reactive with compounds common to all organisms, the decisive factors regulating species specificity are probably the degrees to which they are accumulated in cells and the internal pH of the cells since there is approximately a tenfold decrease in the reactivity of these compounds for each unit drop in pH.

Cycloheximide—possibly the best known of the antifungal antibiotics— is a fungitoxicants that is apparently considerably more specific in its mode of action than are most of the synthetic organic fungicides. However, its selectivity toward species is no better and perhaps much poorer than that of many of the organic fungicides. Sensitivity to this antibiotic is widely distributed among organisms, as is evidenced by its toxicity to higher plants and animals, fungi, and protozoa, and yet it possesses a most remarkable specificity for certain closely related species. For example, growth of some species of yeasts is absolutely prevented by 0.17 p.p.m. of cycloheximide, whereas under the same conditions other species grow even in the presence of 1000 p.p.m. It inhibits germination of conidia of some fungi without any effect on oxygen uptake. Respiration in other species is markedly affected. For example, concentrations that prevent germination of spores of *Myrothecium verrucaria* reduce their oxygen consumption by about 85%. With *F. roseum*, oxygen uptake is depressed about 50% by concentrations that prevent growth, but even much higher concentrations do not significantly increase the effect on respiration. This antibiotic brings about a marked reduction in endogenous respiration and interferes with oxidation of sugars, amino acids, and organic acids by the protozoan *Tetrahymena pyriformis*.

The mode of action of cycloheximide is not presently understood and, while it is known to induce mitotic disturbances in some organisms (as is pointed out later in this chapter), it seems unlikely in view of its effects on metabolism that its toxicity can be accounted for entirely in this way.

Normally, one would expect enzymes or proteins at the outer surface of the cell to be affected by ions of the heavy metals, but they most probably react with labile chemical groups in cellular components where ever they come in contact with them. The lack of specificity of these fungicides is indicated by the fact that they are inhibitory to a wide variety of enzymes. The enzyme they attack in the living cell may depend on its location in the cell. There are essentially no grounds for assuming that the enzymes of pathogen or host cells or of cells of different pathogens would be markedly different in their tolerances to heavy metals such as copper, mercury, and silver. Any tolerance that cells may have for these toxicants would probably rest on the nature of the sensitive groups that are exposed at the surface of the cytoplasm, on their ability to exclude them from the cytoplasm, or on their ability to bind them into nontoxic complexes.

The mechanism of cytotoxic action, defined in terms of interference with a specific biochemical reaction, has not been established conclusively for any of the common fungicides. The problem of establishing a specific basis for cytotoxicity is complicated by several factors. For reasons that have been indicated, the particular reactions or processes that are affected by a given toxicant may not be the same in all susceptible species even though the chemical basis for toxicity is the same. The chemical reactivity of most of the fungicides suggests that they may often affect several processes simultaneously and to about the same degree. Moreover, they may bring about secondary effects that tend to obscure less obvious primary effects.

C. Accumulation of Toxic Metabolites and Enzyme Stimulation or Induction

It is conceivable that a fungitoxicant might interfere with metabolism in such a manner that toxicity results directly from the effects of an accumulated metabolite. While most metabolites would not be harmful except in relatively large quantities, hydrogen peroxide is an example of one that would be detrimental even in small quantities. It has been demonstrated that sublethal doses of cyanide increase the mutation rate in *Neurospora* and this effect is attributed to hydrogen peroxide accumulated as a result of inhibition of catalase by the cyanide. Similar results have been obtained with cells of *E. coli* treated with azide. Lethal effects from peroxide might well result if a fungicide exerted a sufficiently inhibitory effect on catalase.

As opposed to inhibition of enzymes, stimulation of an enzyme into abnormal activity would constitute a mechanism of toxicity. It has been postulated that the effect of dinitrophenol on oxidative phosphorylation may be the result of its activation or stimulation of ATP-ase in the mitochondria.

As opposed to inhibition of enzymes, stimulation of an enzyme into abnormal activity would constitute a mechanism of toxicity. It has been postulated that the effect of dinitrophenol on oxidative phosphorylation may be the result of its activation or stimulation of ATP-ase in the mitochondria.

Perhaps one of the most effective and subtle means of killing a fungus cell would be to induce its enzymatic self-destruction. Prestidge and Pardee (1957) presented evidence that penicillin induces formation of a lytic enzyme in cells of *E. coli* that attacks the cell membrane. Compounds like chloramphenicol and 7-azatryptophan, that inhibit synthesis of enzyme protein, protect cells from penicillin.

It should be recognized that the ultimate cause of death of fungus cells, in which normal metabolism is disrupted by a fungicide, may in many cases be a form of autolysis in which cellular components are irreversibly altered by enzymes.

VI. BIOLOGICAL MODIFICATION OF FUNGICIDES

Susceptibility of an organism to a fungicide may in some instances depend on whether it can activate or detoxify the fungicide. If a compound is modified by an organism its toxicity may be increased, decreased, or remain unaltered. Generally, mechanisms for altering foreign chemicals involve metabolic systems in existence in the cells prior to their exposure to a toxicant, but there are exceptions to this generalization. Penicillinase accounts for the resistance of certain bacteria to penicillin. Development of penicillinase is an adaptive process in some species. Horsfall (1956) cited instances in which appreciable periods of association between fungi and fungicides are required before they are able to overcome the toxic effects, which suggests that detoxification awaits development of detoxifying mechanisms in the cells. Activation or detodxification may involve direct enzymatic action on the toxicant or, in other cases, they may involve combination of the toxicant with metabolites or other cellular components. Conceivably, a toxicant may remain unaltered, but nevertheless be rendered ineffective in a particular biological system by a competitive metabolite the accumulation of which is incited by the toxicant.

A. Activation

Although biological activation has received less attention than detoxification, it may be a significant factor in controlling toxicity and specificity of certain fungicides, insecticides, and growth regulators. Toxicity of organic phosphorus insecticides is generally attributed to interference with cholinesterase activity. Preliminary *in vivo* oxidation is required to convert some members of this group into active cholinesterase inhibitors. Beta oxidation of the side chains of certain phenoxyalkylcarboxylic acids by plant tissues is reported to be necessary for their conversion into active herbicides. Different species of higher plants vary in their ability to bring about this oxidation, which probably accounts for species specificity of these herbicides.

Dichlone (2, 4-dichloro-1, 4-naphthoquinone), the corresponding naphthohydroquinone, and its diacetyl ester are equally toxic when applied to certain fungi. Byrde and Woodcock (1953) presented good evidence that the diacetyl ester is biologically activated by conversion, through hydrolysis, to naphthohydroquinone, and Horsfall (1956) postulated that a phenol oxidase finally converts it to the toxic quinone form. The importance of biological modification of dialkyldithiocarbamate fungicides is not entirely clear, but it is apparent, as is indicated by the following examples, that these fungicides are subject to alteration by fungi. Cells of *F. roseum* decompose thiram with release of CS_2. This gas apparently arises from acid decomposition of dimethyldithiocarbamate ions formed as a result of biological reduction of the thiram. Goksøyr (1955) demonstrated by spectrophotometric means that cells of baker's yeast can reduce thiram, and Johnston (1953) showed that thiuram disulfied can be reduced by glutathione. On the other hand, Keilin and Hartree (1940) found that the cytochrome oxidase system in homogenates of animal cells oxidizes sodium diethyldithiocarbamate to antabuse which strongly inhibits the succinoxidase system, and Neufeld et. al. (1957) prepared an extract from *Piricularia oryzae* that performs the same kind of oxidation. Biological oxidation and reduction of the dialkyldithiocarbamates may possibly be of importance in regulating their toxicity, or such reactions may simply be examples of metabolism of foreign chemicals in which toxicity is not significantly altered.

Sproston (1957) described an interesting case of activation associated with the natural resistance of *Impatiens balsamina* to fungal infection. The fungitoxicants in this case appear to be polyphenolic compounds that occur in the plant cells probably as glycosides or other bound forms. When spores of any of several fungal species germinate on the leaves and the hyphae penetrate the epidermal cells, the invading fungi apparently free or otherwise activate the polyphenols with the result that both the invaded plant cells and the fungus are killed.

B. Detoxification

Inherent detoxification mechanisms may account for the insensitivity of some species of fungi to toxicants that destroy other species. For example, strains of *Endomyces vernalis* that are resistant to pyrithiamin decompose it, whereas sensitive strains do not. Desthiobiotin strongly inhibits growth of *Lactobacillus casei* but actually has biotin activity for a number of yeasts and filamentous fungi. The latter organisms are apparently able to insert sulfur into the molecule thus converting it to biotin.

Quinones and phenols are fungitoxic compounds which may be subject to detoxification by fungal cells. In a study of toxicity of 44 phenols and quinones, Rich and Horsfall (1954) showed that those compounds which were oxidized and polymerized to yield colored products, when exposed to an extract from the mycelium of *Stemphylium sarcinaeforme*, are usually nontoxic to spores of the fungus. Of the compounds that yield a colored product 88% are nontoxic, while of those that do not yield a colored product 84% are toxic. A similar but less striking relationship exists between toxicity and ability of mycelial extracts of *Monilinia fructicola* to oxidize and polymerize quinones and phenols.

Aspergillus niger detoxifies 2-naphthyloxyacetic acid and β-(2-naphthyloxy) propionic acid by hydroxylating them at the 6- position and reduces the toxicity of phenoxyacetic and β-phenoxypropionic by hydroxylating them at either the *para* or *ortho* position on the benzene ring.

Some fungi tolerant of arsenic metabolize it to volatile trimethyl arsine and this may represent a detoxification mechanism, but there are other species with high tolerance that do not metabolize arsenic in this manner.

Certain metabolites already present in fungal cells at the time of treatment undoubtedly contribute to protection of vital cellular components by reacting with fungitoxic compounds. Many amino acids will antidote the toxicity of coppr sulfate to *S. sarcinaeforme* and *M. fructicola*. A number of fungicides interfere with sulfhydryl groups of proteins and coenzymes, and naturally occurring sulfhydryl compounds such as cysteine and glutathione protect the proteins and coenzymes by combining with the toxicant and rendering it ineffective. Sulfhydryl compounds are known to react with heavy metals, with nabam, with captan with -*s*-triazines, and with quinones to form nontoxic derivatives.

Since detoxification often makes use of essential metabolites and of normal metabolic mechinary, it should be expected that depletion of metabolites or overloading of the systems might ultimately result in death of the cells. The case of sulfur is a good example. Its toxicity to fungal cells is lowered when they metabolize it to hydrogen sulfide, but the process of detoxification, i.e., reduction, is thought to be responsible for death of the cells. In animals, cysteine is involved in the detoxification of bromobenzene. Large doses of bromobenzene cause cessation of growth because there is insufficient cysteine for both detoxification and growth. Supplies of cellular components capable of inactivating toxicants are limited and can confer only a limited degree of resistance to a fungicide. However, the relatively large quantities that are required of many common fungicides on a spore weight basis to inhibit germination have been attributed to more or less indiscriminate interaction between toxicant and various cellular components. For instance, at subinhibitory concentrations dichlone reacts, with substances from conidia of *N. sitophila* to form at least five different products. If any of the reactants are vital metabolites, they occur in the spores in sufficient quantities to detoxify appreciable quantities of dichlone with enough remaining to support germination. It is apparent that one of the characteristics of a fungicide of high efficiency (on a microgram-per-gram basis) must be that of discriminate reactivity with cellular components.

Detoxification by reaction with a nonvital cellular component occurs in certain wood-destroying fungi that produce oxalic acid. In these organisms resistance to copper fungicides results from formation of the relatively nontoxic copper oxalate.

VII. ACQUIRED RESISTANCE TO FUNGICIDES

Sustained, acquired resistance to fungicides in populations of plant pathogens has not developed into a problem comparable to that of acquired resistance to antibiotics in bacterial populations or of acquired resistance in insects to certain organic insecticides. Reports of fungicide-resistant strains of plant pathogens in the field are limited. Taylor (1953) collected spores of *Physalospora obtusa* from a number of orchards and found greater tolerane to Bordeaux mixture in those from orchards that had been continuously sprayed with that fungicide than in those from orchards not so sprayed. There is abundant evidence from the laboratory, however, that fungi either have the ability to develop tolerance to diverse kinds of fungicides or else that natural populations contain a few potentially resistant individuals that gain the ascendancy in the population when the organisms are cultivated in the presence of the toxicant. Whiffen (1948) found that *Saccharomyces pastorianus* acquired a sixteenfold increase in tolerance to cycloheximide after three transfers in medium containing the antibiotic, and Roper and Käfer (1957) isolated, from a medium containing acriflavine, three strains of *A. nidulans* that were resistant to this toxicant. In each of the three strains resistance was due to single gene mutation.

There are reports of strains of fungi that are more tolerant than the normal population to arsenic, to thiram and organic mercury, to tetrachloronitrobenzene, and to copper sulfate. In some of these cases the fungi retained their tolerance when cultivated in the absence of the fungicide, which indicates a permanent change, whereas others lost their tolerance. Leben et. al. (1955) exposed conidia of *Venturia iaequalis* to ultraviolet light, but did not find among the survivors mutants that were resistant to thiram or to sulfur.

A facor that may account, in part, for failure of resistant strains of pathogens to develop in the field may be the nature of the environment in which spores of most plant pathogens are exposed to fungicides. Leaf surfaces are relatively poor in substrates and there is good evidence that resistant strains are less likely to develop when certain nutrients are in limited supply or are unavailable. This is suggested by the more frequent reports of resistance arising in culture medium in the laboratory than in the field. Sevag and Rosanoff (1952) noted that when either aspartic acid or phenylalanine are absent from a medium that contained steptomycin, resistant strains of *Micrococcus* fail to emerge from the population, although strains of the bacterium, which have developed resistance in the presence of the these amino acids, retain their resistance to streptomycin when grown in medium lacking aspartic acid and phenylalanine. Schnitzer and Grunbert (1957) discuss other instances in which emergence of forms of bacteria that are resistant to toxicants is dependent on the composition of the medium. Mechanisms of resistance that involve utilization of detoxifying metabolites or changes in pH may be practically inoperative in an environment of limited exogenous substrate.

Another, and likely more important, factor in preventing emergence of fungicide-resistant forms in the field is the less specific mode of action of most fungicides as compared to those of the antibiotics and of some of the insectides. When a toxicant can affect a number of vital processes, any one of which may inhibit growth, there is less likelihood that resistant strains of an organism will develop than when toxicity is based on more specific effects. If development of resistance is dependent on gene mutation, one would expect that strains of pathogens resistant to mutagenic fungicides would be more prevalent than those resistant to fungicides which are not mutagens, but apparently this possibility has never been investigated.

As fungicides that are more complex chemically and more specific in their modes of action come into general use, the problem of resistance of pathogens to fungicides in the field is likely to arise.

VIII. MUTATION, MITOTIC DISTURBANCES, AND MORPHOLOGICAL ABERRATION

A. Mutagenesis

A number of fungitoxic compounds produce mutations or affect the process of mitosis. In considering the physiology of fungitoxicity one should be concerned with mutagenic effects both because they may be involved in the mechanism of action of the fungicides and because of any possible hazard that they may constitute for people who handle or who are otherwise exposed to such toxicants. Mutagenic chemicals are usually recognized by the heritable abnormalities that they produce in the progeny of those cells that survive treatment. When cells do not survive treatment with a fungicide, the cause of death might well be lethal mutation without being recognized as such.

In addition to gene mutations, there may be cytological mutations involving structural changes in chromosomes such as breaks, deletions, and inversions. Inasmuch as chromosomes of most fungi are too small for easy cytological study, much of the evidence for fungicidal effects on chromosomes and mitotic behavior is indirect and is based on observable effects in root tips of *Vicia faba* and *Allium cepa*. Toxic chemicals sometimes cause pycnosis or stickiness of chromosomes. In the most severe instances, the chromosomes appear to melt together and recovery of cells so affected is impossible. Some chromosomal aberrations may not arise from direct action of a toxicant, but rather may represent post lethal autolytic damage, possibly involving the action of deoxyribonuclease or of proteolytic enzymes.

Quinones, phenols, and 8-hydroxyquinoline are among the fungitoxic compounds reported by Levan (1951) to be mutagenic in root tip tests with higher plants. The fungicide, tetrachloronitrobenzene, induces mutations in *F. coeruleum*. Fries and Kihlman (1948) produced mutations in *Ophiostoma multiannulatum* with certain methyl xanthines. The mutations produced by chemicals of this type may involve interference with nucleic acid metabolism. It seems possible that these purine analogues may replace normal components and result in the synthesis of defective nucleic acids. This would represent a case of lethal synthesis for those cells that do not survive. Mutagenesis may result from accumulation of toxic metabolites. Compounds that poison catalase, which normally destroys hydrogen peroxide in cells, bring about an increase in the rate of mutation in *Neurospora* and in bacteria.

Some of the same fungitoxicants that are known to induce lethal mutations, drastic structural changes in chromosomes, or abnormal chromosomal behavior are capable also of other effects in cells. In such cases, it may be difficult to determine which effect is primarily responsible for their general toxicity to fungal populations. An excellent example is presented by the nitrogen mustards which are well-known mutagens, but which also inactivate the enzymes, hexokinase, adenosine triphosphatase, choline, oxidase, and acetylase. Herriott (1948) found, however, that sulfur mustards inactivate viruses, especially those that contain deoxyribonucleic acid, more readily than they inactivate enzymes. He suggested that this was the result of reaction with nucleic acid. If this is true also of the nitrogen mustards, it consitutes strong circumstantial evidence that their primary toxic effects on cells may involve the chromosomes. Formaldehyde causes mutations in *Neurospora* as well as in higher plants, bacteria, and *Drosophila*. Its wide use in inactivating viruses is ample evidence of its activity against nucleoproteins. Cartwright et. al. (1956) suggested a number of possible sites of lethal action of formaldehyde with the tobacco mosaic virus. Included are the amino groups of the purines and pyrimidines and the hydroxy groups of ribose in the nucleic acid. These appear to be the most likely sites of action, because it is generally conceded that the nucleic acid portion of the virus contains the biologically active centers. Viruses and chromosomes are similar in their chemical composition. Thus with formaldehyde, as with the mustards, there is reason to suspect that the primary lethal effect may be on the chromosomes, but this need not necessarily be the case because a number of vital components other than nucleic acids of the chromosomes are reactive with formaldehyde and in many cases may be more accessible to the toxicant.

B. Interference with the Mitotic Spindle

In the process of cell division the orderly movement of the two sets of chromosomes to opposite poles of the dividing cell involves the formation and functioning of the mitotic spindle. Appearnance of the spindle mechanism in some cells is accompanied by abrupt changes in the viscosity of the protoplasm—sometimes referred to as mitotic gelation—and the spindle fibers appear to crystallize from the protoplasmic fluid. They are similar in appearance to tactoids or liquid crystals and their structure may involve hydrophobic linkages between proteins or between proteins and lipids. Malformation or malfunction of the spindle results in polyploidy, unequal distribution of nuclear material, or other cytological aberrations. Such phenomena are sometimes initiated by fungicides. In general, fungicides that affect development or function of the mitotic spindle differ from those that affect chromosomes directly.

One major group of compounds that affect spindle formation is the lipid-soluble compounds whose activity is correlated with their physical properties rather than with their chemical reactivity. Included in this group are naphthalene and benzene and their derivative and most probably many other compounds of this type. Fat-soluble anesthetics and colchicine, at concentrations that prevent cell division, typically prevent mitotic gelation in sea urchin eggs. Compounds of this kind probably exert their effects through interference with hydrophobic linkages in the spindle structure.

Organic mercury compounds constitute a second group with rather high activity in distrupting spindle function, but they probably act by a different mechanism from the previous group of compounds. Levan (1951) reported that butyl mercury iodide is about 1,000 times more effective than is colchicine in inducing c-mitosis in onion root tips. Organic mercury fungicides, used as treatments on rye seed, have been known to induce tetraploidy.

Cycloheximide consistently produces aberrant mitotic behavior in onion root tips and affects meiosis in *Gymnosporangium* in a manner suggesting malfunctioning of the spindle. A list of compounds that are reported to produce mitotic aberrations has been compiled by Horsfall (1956).

C. Effects on Morphology

Morphological abnormalities, such as giant cells, hyphal monstrosities, and a filamentous habit in organisms that normally grow by fragmentation into individual cells, are among the effects produced in plant pathogens, other fungi, and bacteria by certain toxic agents. Responses of this kind imply interference of such a nature that the cell, although unable to divide normally, retains its capacity for growth—at least for a while. Certain specific inhibitors of cell division characteristically induce bacteria to grow in filamentous from, and yeast to produce enlarged paired cells.

Although general interference with mitosis may in some cases account for abnormalities of the kind under discussion here, there are other instances in which there is a more specific effect, posibly involving dissociation of the processes of nuclear division from cleavage of the cytoplasm or of the cell wall. Nickerson (1948) advanced the hypothesis that cell division normally involves a chain of enzymatic reactions that can be uncoupled from growth processes. It is thought that sulfhydryl containing substances play a role in fragmentation, because such compounds as cysteine and glutathione promote cell division and antagonize the effects of such substances as cobalt and penicillin that tend to prevent cell division.

Horsfall (1956) has complied and discussed much of the literature on toxicants that incite morphological abnormalities in fungi.

Allison and Christensen (1957) described the injurious effects of the antibiotic, filipin, on *Helminthosporium sativum* as including mutations, tumorlike growths, and rupturing of the cells with extrusion of the protoplasts. Among the chemicals that inhibits cell division in *E. coli* without inhibiting growth are: acriflavine, diazouracil, diethyl sulfate, ethylenimine, 5-nitro-7-hydroxybenzothiazole, and β-propiolactone. Nitrogen mustard and triethylenemelamine specifically inhibit cell division in baker's yeast. Morphogenic effects are also produced by appropriate doses of irradiation. Ultraviolet treatment of *E. coli* results in this developing in the form of "long filaments" or "snakes." The effect appears to be quite specific since syntheses of deoxyribonucleic acid, ribonucleic acid, and proteins continue at the same rates in irradiated as in unirradiated cells. At proper levels, α–, γ–, and X-irradiation also inhibit cell division without affecting growth of *E. coli* and of *S. cerevisiae*. It is interesting that many of the agents that specifically inhibit cell division are also mutagenic, but there is no definite evidence that the failure of cells to divide after treatment with these agents is a result of damage to chromosomes. Death ultimately results if cells fail to regain their ability to divide normally.

X. SPORES VERSUS MYCELIUM

A. Effects on Sporulation

Sporulation in filamentous fungi is a fragmentation process that apparently does not differ in any fundamental way from the behavior of bacteria and yeasts that normally reproduce by fragmentation into individual cells. The process may be considered as consisting of two steps—mitosis in which nuclear material is distributed properly between the mother and daughter cells and, second, cell division involving cleavage of the cytoplasm and organization of new cell wall material in such a way that the spore separates from the structure on which it is produced.

A number of fungus diseases of plants could be controlled effectively if sporulation could be prevented so as to reduce the inoculum potential. Horsfall and Rich (1955) became interested in compounds that might selectively inhibit sporulation as a possible means of plant disease control. They compared a number of compounds, including mitotic poisons, chelating agents, and others as to their relative effectiveness in reducing mycelial growth and in inhibiting production of cônidia by *Monilinia fructicola.* Some of the mitotic poisons showed greater inhibitory activity toward sporulation than toward mycelial growth, as did also a number of the chelating agents. Several compounds, including ketones, acids, and amides that are α–beta unsaturated, actually enhanced sporulation, while among the compounds that were most active in selectively suppressing sporulation in these experiments were sodium thiocyanate, benzidine hydrochloride, diphenylthiocarbazone, and bis (4-dimethylaminophenyl) methane. Metals may promote sporulation (fragmentation) in some instances, but in others have the opposite effect. Cobalt, for example, induces certain yeasts to grow in filamentous form. Zinc is involved in the fragmentation process in cultures of *Ustilago sphaerogena*, but the exact nature of involvement is not understood. Compounds that react with metals might logically be expected to affect sporulation and Rich and Horsfall (1948) found that the metal chelator, dimethylglyoxime, prevents sporulation in *A. niger* at concentrations that do not appreciably affect growth of mycelium. Inhibition of sporulation in *A. niger* at concentrations that do not appreciably affect growth of mycelium. Inhibition of sporulation in *A. niger* by thiourea and in *Peronospora destructor* by limesulfur is considered to result from interference with essential metals.

It is well known that sporulation of many plant pathogenic fungi in culture is sensitive to minor nutritional and environmental changes that do not appreciably affect mycelial development. If sporulation in nature is likewise a more sensitive process than growth, it would appear to be appropriate to continue and to intensify the search for specific chemical inhibitors of sporulation, because reduction of inoculum potential might constitute an easier and more efficient means of controlling some plant diseases than attempting to suppress spore germination and infection by use of protectant fungicides.

B. Comparative Sensitivity of Spores and Mycelium

Some toxicants are more effective in inhibiting spore germination than they are in inhibiting mycelial growth, whereas other toxicants exhibit the opposite effect. Diphenyl, for example, does not inhibit germination of spores of several species of fungi, but it is inhibitory to hyphal growth of these organisms. Spores of the fungus *Trichophyton purpureum* germinate in solutions containing sufficient sulfanilamide to completely inhibit hyphal growth. Horsfall and Rich (1953) reported a number of compounds that inhibit mycelial growth of *M. fructicola*, but fail to prevent spore germination.

Since germination does not necessarily involve nuclear division or require the transport of substrates into the cell, as in the case in mycelial growth, it has been proposed that interference with these processes could logically account for the greater sensitivity of hyphae. Spores ordinarily carry greater reserves of certain vital metabolites than mycelium and interference with the synthesis of these would not become apparent during germination, but would be immediately apparent in hyphal growth.

Higher sensitivity of spores than of mycelium, may be explained in some cases by the relatively slow rate at which mycleium accumulates toxicants from the medium. However, they showed that on the basis of the ultimate quantities of toxicant taken up, spores and mycelium are about equally sensitive. When interpreting observed differences in sensitivity between spores and mycelium, one must consider whether the amount of fungal material was equivalent in the two cases.

A criticism that might be leveled at the spore germination test for evaluating fungicides is that it fails to detect toxicants which inhibit mycelial growth without affecting germination of spores.

Considerable progress has been made in identifying the cellular activities that are involved when plant pathogenic organisms are controlled by fungicides and in recognizing the factors governing sensitivity and resistance to toxic compounds. Our knowledge of cellular physiology and structure is still too incomplete to permit following a purely rational approach to the design of a specific toxicant for a specific use. For some time progress will probably continue largely along empirical lines, but as our understanding of comparative physiology and biochemistry and of the subtle details of differences at the cellular level between species improves, we will move in the direction of a sounder scietific approach to control of plant diseases by chemical means.

8

CHEMICAL CHARACTERISTICS OF FUNGIOTOXICANTS

This chapter will deal with the intrinsic properties of compounds that make them fungitoxic.

As is the case of almost any discussion of chemical properties relating to biological action, there will be considerable simplification. Killing of living cells by a chemical compounds is not a simple process, but rather the total observable response to a single or multiple chain reaction. Whether or not a specific property of a compound will result in the death of a specific cell or tissue very often depends on an exquisite interaction between the physical-chemical properties of the compound and those of the cell affected. For example, fluoroacetic acid is produced in the poisonous plant *Dichapetalum* without harm to the plant. Peters (1952) has found that when mammals ingest the plant, the fluoroacetate is mistaken for acetate in the Krebs cycle and fluoroacetate is produced which inhibits aconitase. This causes citrate to accumulate. Peters feels that the excess citrate is the toxicant. Here is a delicate change with a profound effect. Fluoroacetate is not a direct poison, but may be changed by mammalian tissue into another compound, fluorocitrate, which causes the accumulation of the toxicant. It would indeed be an over simplification to ascribe to one property of fluoroacetate the ability to kill directly.

What must a compound be able to do in order to poison a cell? Simply stated, a poisonous compound must be able to get to a site of action, resist changes in transit, and exert a deleterious effect when it arrives. Requirements for poisonous molecules will, therefore, differ according to where they exert their effect. If they act outside the cell or at its surface, they will have no need to penetrate the intricate cellular membranes. In this case, the requirements for toxicants would be much simpler and less specific than for compounds which must penetrate.

Most organic molecules enter living cells most efficiently in the nonionized state. Organic toxicants, therefore, are usually most toxic in their nonionized form.

Are molecules always most toxic in their nonionized form? Albert (1951) and his group have found that the bactericidal effectiveness of the acridines and related series depends on their ability to ionize. Also, the addition of an aliphatic side chain, which usually aids permeation and attendant toxicity, actually reduces the bacteritoxicity of these compounds. Albert has deduced that acridine must act at the surface of the bacterial cells.

The 8-quinolinol derivatives, on the other hand, must be nonionized for maximum effect. If, as in the case of 8-quinolinols, the toxic molecule must enter the turbulent laboratory of the living cell, the molecule must resist destruction before it reaches its site of action. Rich have demonstrated correlation between the detoxifying enzymes present in fungi and the resistance of these fungi to poisoning.

What does this mean in terms of comparative reactivity? A toxic molecule which is highly reactive and relatively unstable would be most likely to act in the outer portions of the cell. Great reactivity might serve to destroy and detoxify such a molecule before it moved very deeply into the cell. This property may, in fact, destroy a compound in the ambient fluid before it reaches the cell. For example, the presence of cysteine in the ambient fluid destroys captan before it can poison yeast cells. Lukens and Rich find that if yeast cells are treated with cobalt and then washed free of excess cobalt, they are more susceptible to captan than cells not treated with cobalt. Assuming that cobalt ties up surface sulfhydryl groups, the pretreatment presumably prevenis the partial breakdown of captan on its way to its of action within the cell.

A more stable, relatively unreactive molecule would have a greater chance of penetrating unchanged to a site of action deeper within the cell. It will be seen that compounds considered chemically inert may be toxic.

Not only the relative site of the action, but also, the mode of action dictates requirements for a toxic molecule. For example, a competitive antimetabolite has more and subtler requirements than those of a nonspecific poison, such as phenol. Sulfanilamide must not only be able to penetrate to a site of action and be resistant to detoxication, but in addition its molecular configuration must very closely approximate the metabolite *p*-aminobenzoic acid with which it competes within the cell. The distances between atoms in the molecule and their spatial relations are critical to the action of sulfanilamide.

Much has been said of the solubility of toxicants in relation to their toxicity. Does a compound have to be water-soluble to be toxic? How else can it be reactive in the water-flooded interior of the cell? Actually, such materials as copper oxide, sulfur, and dichlone, highly insoluble in water, are toxic. Very often water solubility is inversely related to toxicity as in the analogues of metallic 8-quinolionates. This does not imply that good solubility in water precludes toxicity. The organic mercury salts, highly water-soluble, are excellent fungicides. Usually those toxicants which are water-soluble, are very reactive compounds exerting general nonspecific toxicity.

Why this wide variation in the relation between toxicity and water solubility? As the sites of action may be in polar or nonpolar milieu, or in a medium with both polar and nonpolar groups, e.g., proteins, toxic compounds must and do vary in their solubilities depending on where they must go to become toxic.

Most often in the case of organic toxicants, of primary importance is not water solubility or lipid (nonpolar) solubility, but rather the partitioning of the compound between the two phases. Rich and Horsfall (1952) state the case for the importance of this relationship in the action of fungicides. Working with a series of nitrosopyrazole analogues, they showed that increasing the oil-water partition coefficient from 0.48 to 6.66 increases fungitoxicity ten thousandfold with no appreciable change in the dosage response slopes of the analogues. They interpret this to mean that the change in the direction of high oil-water partition has greatly increased the penetration of the compound into the cell without altering the mode of action.

This solubility balance may be so critical that it may decide whether a compound will be toxic to one plant species and not to another. Wellman and McCallan (1946) working with 2-substituted imidazoline derivatives found that the tridecyl derivative was fourteen times more phytotoxin than the heptadecyl derivative. The heptadecyl homologue however, was twice as toxic to fungi as the tridecyl homologue.

A more complicated relationship between toxicity and comparative solubilities exists in the 8-quinolinol derivatives. Albert and Hampton (1954) and Albert et. al. (1954) showed that the antibacterial action of oxine derivatives in directly and closely correlated with oil-water partition. Albert et. al. (1953), with bacterial, and Block (1956) and Sijpesteijn et. al. (1957), with fungi, have demonstrated by various indirect biological methods that the toxic species of oxine and its derivative is the water-soluble half chelate. When the system is flooded with oxine well in excess of that needed for full chelation, there is no toxicity. Excess oxine would maintain the full chelate within the cell, preventing it from going to the half chelate. Thus, high oil-water partition is essential for movement of the molecule to a site of action, but, once there, the water soluble half chelate must be present for toxicity. This does not imply that water solubility per se is essential to the toxicity of the half chelate but merely that high oil-water partition is essential to movement, and once inside the cell this condition is not essential to toxicity.

The toxicity of another group of compounds appears to be completely correlated with oil-water partition, with no evidence that any other property is involved. These are the so-called inert, nonspecific toxicants which follow the Ferguson effect, to be discussed later.

The properties aiding direct permeation into a cell may not be the only ones of importance in moving a damaging number of toxic molecules to the site of action. Sussman (1954) showed that dormant ascospores of *Neurospora* adsorb ethylenediamine tetraacetic acid to their surface. The adsorption takes place even if the spores are first killed by boiling. If the ethylenediamine tetraacetic acid :s removed from the spores before their dormancy is broken, there is not toxicity, but ethylenediamine tetraacetic acid is toxic if present when germination processes begin. Sussman interprets this to mean that the ethylenediamine tetraacetic acid cannot enter the dormant spore. Not until the permeability properties of the spores change by the onset of germination can the toxic molecules enter. Lowry et. al. (1957) presented evidence that metallic cations such as Ag^+, $Cu^{+}+$, and $UO_2^{+}+$, follow this same course. Here, then, is an example of adsorptive forces concentrating a toxicant at the surface of a cell, ready to enter when the permeability properties of the cell change.

The work of Sussman and his colleagues should not be taken to mean that metallic ions enter only in the fashion they describe. Miller et. al. (1953) have shown that fungus spores take up silver and cerium extremely rapidly and greatly concentrate the toxicants in so doing. With silver, for example, as high as 2124 p.p.m. of spore weight may be taken up by *Monilinia fructicola* spores in the first 30 seconds of exposure to a maximum possible dose of 4160 p.p.m. Cerium entered *Aspergillus niger* spores at the rate of 7625 p.p.m. of spore weight in the first 30 seconds of exposure to a maximum possible dose of 10,000 p.p.m. *Miller* et. al. (1953) point out that if doses of this magnitude were the result of simple adsorption, then the spore surface would have to be covered with multimolecular layers of toxicant. This does not appear likely.

Owens and Miller (1957) later studied the distribution of cerium and silver toxicants in spores. These workers disintegrated toxicant-treated spores by sonic vibration, then fractionated the spore fragments by centrifugation. They found that only 4% of the total amount of cerium or silver taken up was associated with the cell walls. Owens and Miller feel that adsorption of cerium or silver on spore walls or plasma membranes plays little or no role in the movement of these toxic ions into the conidia of their two test organisms, *Neurospora sitophila* and *Aspergillus niger*.

Having the properties necessary to move to site of action without being detoxified, how may compounds poison? The following two sections discuss specific inorganic and organic toxicants in detail. The last section will deal with fashioning fungicides and with the future of fungicides.

I. DESTRUCTIVE PROPERTIES OF FUNGIOTOXICANTS

Very few fungitoxic mechanisms are fully understood. Possession of one of the properties discussed below does not necessarily make the molecule toxic, nor does it explain why the molecule is toxic. Rather, this is a discussion of chemical and physical properties by which molecule may disrupt essential life processes.

A. Direct Chemical Interactions

Toxicants may undergo direct chemical interactions with vital constituents of the cell. For example, irreversible addition or substitution reactions with important enzymes or substrates would effectively poison a cell. Even reversible reactions will inhibit manifestations of life, and eventually cause death if a continuing supply of the toxicant is fed to the cell. These reversible reactions, inhibiting only as long as the toxicant is present, exemplify fungistasis or bacteriostasis.

Whether or not a toxic reaction is reversible, will depend largely on the type of bonds formed between toxicant and vital component. Albert (1951) summarizes bonds of importance in biology and gives their usual bond strengths. The *van der Waals' bonds* is a weak bond (about 0.5 kilocalories) which may occur when any two atoms or two different molecules approach each other. Atoms must be able to approach each other very closely if they are to be firmly bound by van der Waals' bonds. Although a single one of these bonds is weak, there may be many of them between molecules of toxicants and vital constituents. The formation of many van der Waals' bonds is more likely to occur between molecules which can fit close together because of their molecular configuration.

The hydrogen bond, with a strength of 2 to 5 kilocalories, is fairly common in biology. In biological systems the hydrogen bonds donors are usually the groups ≡N⁺—H, =N—H, or —O—H. The hydrogen bond accepts oxygen or nitrogen atoms, or sometimes chloride ions.

The hydrogen bonds is apparently made by resonance between two structures where hydrogen is attached to one or the other of the atoms which it joins. There is supposedly no actual oscillation, but the distribution of electrons is intermediate between the two forms [A—HB] and [AH—B] and is represented by [A—B]. The hydrogen bond is stable, because the energy of the mesomeric state is less than that of either of the other structures.

Ionic bonds, with a usual strength of about 5 kilocalories, form between oppositely charged ions. This is the bond of salt formation.

By far the strongest is the *covalent bond* with bond strengths of from 40 to 100 kilocalories. This bond, in which combining atoms each contribute one shared electron, makes for essentially irreversible combinations between toxic molecules and vital components. Such reactions as alkylation, acylation, or halogenation are most likely to cause irreversible poisoning. For example, Burchfield and Storrs (1956) suggest that 2, 4-dichloro-6 (*o*-chloroanilino)-*s*-triazine may be so effective as a fungitoxicant because it alkylates amino acids.

Factors other than bond strength are also important in toxicity by direct chemical interaction. One of these factors is reaction rate. Burchfield and Storrs (1957) studied the importance of reaction rate in the activity of toxicants. Specifically, they investigated the kinetics of reaction *in vitro* between toxicants and vital constituents of cells. Although two toxicants may possibly react with the same group of a vital constituent, the ease and speed with which the reaction occurs may well decide whether the reaction is important to the toxicity of one compounds and not the other. Some of the factors discussed were energies of activation, frequency factors, degree of salvation of a reactive halogen, and the degree of dissociation of the substrate. 1-Fluoro-2, 4-dinitrobenzene was found to be a more highly specific reagent for sulfhydryl groups than is 2, 4-dichloro-6 (*o*- chloroanilino)-*s*-trazine. Conversely, the *s*-triazine is highly reactive with organic amines and phenols, while the 1-fluoro-2, 4-dinitrobenzene is not. This latter difference is apparently related to the low dissociation constants of organic amines and phenols important to the growth of the organism.

Lukens and Siler (1958) suggest that the toxicity of captan is related to double decomposition reaction of captan with sulfhydryl containing components of the cell. Here again, whether or not a compound will react in a certain way is not enough to make it a toxicant. The rate of reaction must be sufficiently fast to be effective. For example, if captan is toxic because of the reaction of —S—$C(Cl)_3$ with essential sulfhydryl groups, then the rest of the molecule may decide the toxicity of the compound as a whole. The phthalimide moiety of captan has carbonyl groups on each side of the *N*. These electron attracting groups weaken the N—S bond. If these electronegative groups were not present, or not conjugation with the N, the resulting N—S linkage would be much more stable. The reaction rates would be slowed down and toxicity would be lessened. For example, our data show that *N*-trichlorome thylthiocarbazole is much more stable than captan and also practically nonfungitoxic.

The heavy metals, such as silver and mercury, combine readily and firmly with sulfhydryl groups. Without question, this property contributes to their toxicity. Certain well-known specific reagents for sulfhydryl groups, such as iodacetamide, are also toxic.

Not only are metals toxic because they may combine readily and firmly with cellular constituents, but also other molecules may be toxic because they combine with metals which are necessary to life. This latter mechanism has been proposed for chelating agents. Binding metals is presumably the action of azide and cyanide.

Bisulfite, toxic to many organisms, is known to combine with pyruvate. This combination would inhibit the decarboxylation of pyruvate, essential to organic acid metabolism.

The α, β-unsaturated ketones, including quinones, form addition products with thiols and amino compounds. This property has been suggested as a basis for toxicity.

There are, then, a number of direct reactions between foreign molecules and cellular components which may lead to toxicity.

B. Effects on Enzyme Systems

One of the most popular areas of research on toxicants has been the study of their effects on enzyme systems. The primary reason for this has been the need to explain the profound effects produced by the application of extremely low dosages of toxicants. A most reasonable explanation is the assumptions that the toxicants hit the organisms in a vital link in their metabolic chain. The many enzymes needed by organisms to change raw food materials into energy and living substance loom large as potential targets. These organic catalysts are not only themselves susceptible to inactivation, but also the processes they perform may be so altered as to result in toxicity. When these effects are described, therefore, they are presented in terms of effects on enzyme systems.

1. *Enzyme Inhibition*. The three principal kinds of inhibition of enzyme systems are termed competitive, noncompetitive, and uncompetitive.

Competitive inhibitors presumably poison by so closely resembling vital substrate that they are taken up by the enzymes which normally work on these substrates. The inhibitor molecules, however, differ sufficiently from the normal substrate molecules so that they cannot be changed by the enzymes into neutrilites needed by the cell to survive. The insertion of these ill-fitting keys into the active sites on the enzyme exclude the required substrate. The competition between the essential molecules and the imposter molecules for enzyme sites is in proportion to the ratio of the two competing molecules in the system. In other words, the degree of enzyme inhibition is proportional to the ratio of substrate to inhibitor. The more substrate, the more inhibitor required for the same degree of inhibition and vice versa.

The classic Lineweaver-Burk (1934) plots of enzyme kinetics compare the reciprocal of substrate concentration S to the reciprocal of the observed reaction rate v. Figure 8.1 represents a typical plot for a case of competitive inhibition.

The slope V of the inhibited line (a) is much steeper than the slope of the noninhibited line (b). The intercepts with the ordinate, however, are identical.

Equation 1 describes the steady state for competitive inhibition. At steady state V = maximum velocity when the enzyme-substrate combination is maximum.

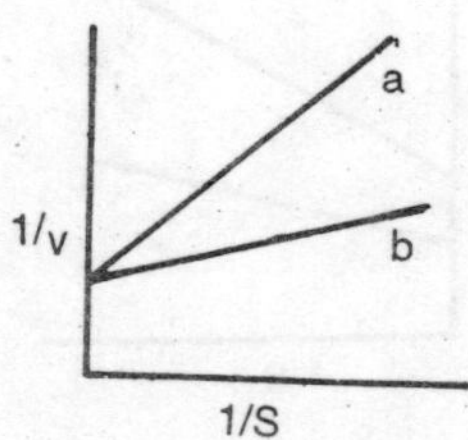

Fig. 8.1. Lineweaver-Burk plot for competitive inhibition.

$$v = \frac{V}{1 + K_s\,[I + (I)/K_i]/S} \qquad \text{(Eq. 1)}$$

K_s = Michaelis constant, which is the substrate concentration at which half maximum velocity is reached.

I = Concentration of inhibitor.

S = Concentration of substrate.

K_i = Constant for dissociation of enzyme-inhibitor combination into enzyme and inhibitor.

As S is increased it affects both K_s and V. When S is sufficiently large, inhibition is completely overcome, and the intercept at sufficient value of S becomes identical with that of no inhibitor. The slopes of the two systems, one with and the other without inhibitor, differ by the function $[I + (I)/K_i]$.

Noncompetitive inhibition applies to the case where the inhibitor combines with the enzyme at a point other than the site at which the substrate attaches. There are two other requirements for this type of inhibition: (*a*) the inhibitor exerts a total inhibitory effect; and (*b*) the binding of inhibitor to enzyme is independent of substrate. Under these conditions, the Lineweaver-Burk plot shows lines differing in both slope and intercept (Fig. 8.2)

Equation 2 describes such a system.

$$v = \frac{V}{[I + K_s/S]\,[I + (I)/K_i]} \qquad \text{(Eq. 2)}$$

Here, as in the preceding equation, the slope of V is still influenced by the inhibitor function, so that the slope of the inhibited system varies from the slope of the noninhibited system by $[I + (I)/K_i]$. It differs, however, by not having the inhibitor function modified by substrate

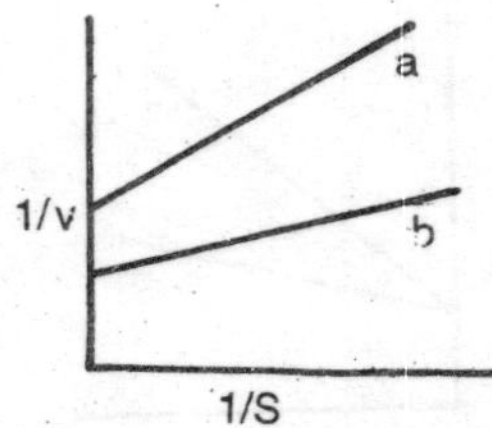

Fig. 8.2. Lineweaver-Burk plot for noncompetitive inhibition.

concentration. Hence, at the intercept, the lines are separated by $[I + (I)/K_i]$. In other words, the reaction rate of a system non-competitively inhibited can never attain the rate of the noninhibited system, regardless of the amount of substrate added.

Uncompetitive inhibition requires that the inhibitor combine only with the enzyme-substrate complex. In these circumstances, the Lineweaver-Burk plot shows parallel slopes for the two systems, and of necessity, different intercepts. Equation 3 describes uncompetitive inhibition.

$$v = \frac{V/[I + (I)/K_i]}{I + K_s[I + (I)/K_i]/S} \qquad \text{(Eq. 3)}$$

Other special cases include inhibition by a partial inhibitor, and inhibition by one of the products of the normal substrate with enzyme.

What does all this mean in terms of *in vivo* toxic effects? The great interest in competitive inhibitors as a toxic mechanism stems from the classic research of Wooks (1940), who demonstrated that sulfanilamide is a competitive inhibitor of *p*-aminobenzoic acid. Not until after Woods' report was *p*-aminobenzoic acid found to be an essential metabolite. Since that time, whenever a toxicant is found which bears a structural resemblance to a vital substrate, competitive inhibition is invoked.

Actually, proven cases of competitive inhibition as a fungitoxic mechanism are not too common. West and Wolf (1955) present strong evidence that 2-heptadecyl-2-imidazoline is a competitive antagonist for guanine and xanthine in *Sclerotinia fructicola*.

Because a compound may act as a competitive inhibitor on one organism is no reason to believe that it acts as such in another. Usually, competitive inhibition is most striking in those cases where the organism is heterotrophic for the particular metabolite being antagonized.

A very striking example of the complexity of enzyme inhibition as a toxic mechanism is described by Woolley and Shaw (1951). They found that 2-azaadenine is particularly lethal to *Lactobacillus* species. They tested the purines, hypoxanthine, adenine, and xanthine as possible metabolites being antagonized, with the following results. 2-Azaadenine competitively inhibits hypoxanthine (Fig. 8.3). Its toxicity is largely

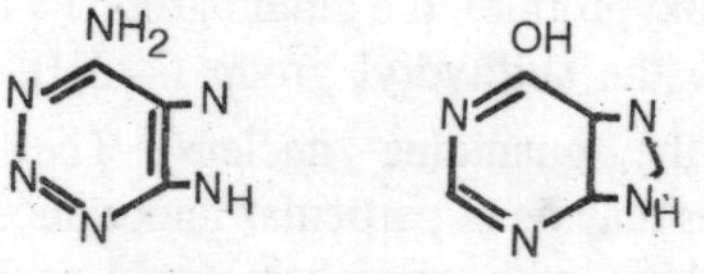

2- AZAADENINE HYPOXANTHINE

Fig. 8.3. Hypoxanthine (right) and its antimetabolite, 2-Azaadenine (left).

overcome by trace amounts of adenine, but is not appreciably alleviated by large amounts of xanthine. Without the inhibitors, xanthine is as effective as hypoxanthine in the purine nutrition of these bacteria. From their results, it becomes obvious that 2-azaadenine is highly specific in its effects on *Lactobacilli*. It is not simply a competitive inhibitor of purines as a class, but of one purine in particular, hypoxanthine.

Their experiment again serves to illustrate the physiological implications that may result from toxicological studies. Their results indicate that the chain of nucleoside synthesis in *Lactobacillus* goes from xanthine → hypoxanthine → adenine. If 2-azaadenine breaks the chain at hypoxanthine, the addition of more xanthine could have little effect. The addition of adenine, however, would be beyond the block and nucleic acid synthesis could continue.

Noncompetitive inhibition as a toxic mechanism may be postulated for almost an toxicant which can denature proteins. For example, picric acid, a protein precipitant, is quite fungitoxic and may act by denaturing proteins. The same may be said for formaldehyde. Any of the amino acid reagents shown to be fungitoxic by Horsfall and Zentmyer (1944) may be acting as noncompetitive inhibitors.

Uncompetitive inhibition has not been postulated as a toxic mechanism for any one particular known toxicant. This type inhibition must surely contribute to the toxicity of some poisons.

Many known fungitoxicant probably act as nonspecific enzyme inhibitors. The enzyme proteins may lose their activity if the attacking toxic molecule has a particular avidity for, or ability to react with any of the reactive groups on the protein.

The susceptibility of enzymes may be realized when we consider the number of reactive sites possessed by amino acids which make up the proteins. Some of these reactive sites are primary amino groups (—NH_2), alcohol groups intermediate between aliphatic and aromatic (the —OH of hydroxyproline), the amino group (= NH), the acid amide group (—$CONH_2$), the sulfhydryl group (—SH), the α-hydrogen of tryptophan, and the guanidine nucleus. The —OH groups of hydroxyamino acids may be of particular importance as susceptible sites because they combine with phosphoric acid to bridge polypetptide chains.

There are other toxic mechanisms closely related to enzyme inhibition which may well be discussed here. One of these, a variation on the theme of competitive inhibition, is a phenomenon classified by Markham (1958) as a type of "lethal synthesis." In this particular mechanisms, the imposter molecules are so like the normal substrate molecules that the toxicants move along the chain of synthesis and are actually incorporated into the final building blocks of living substance. The building blocks containing these imposter molecules are defective and cannot be utilized in a normal fashion by the organism. The organism shows toxic effects as a consequence.

Heinrich et. al. (1952) found that the protozoan *Tetrahymena geleii*, which is readily poisoned by 8-azaguanine, incorporates this aminopurine into its ribonucleic acid when grown on sublethal doses of the toxicant. The toxic effect of 8-azaguanine is readily overcome by the simultaneous addition of the nucleoside, guanosine.

Another example of this type is reported by Dunn and Smith (1957). They found that *Escherichia coli* made sensitive to poisoning by 5-bromouracil incorporates this halogenated pyrimidine into its deoxyribonucleic acid. The toxic effect of this compound cannot be overcome by the addition of thymine for which it is known to be an antimetabolite. *E. coli* is not inhibited by 5-bromouracil nor does it incorporate it unless the bacteria are made to require thymine by treatment with sulfanilamide.

To call this type of toxicity a form of competitive inhibition may not be strictly proper, as it may or may not be reversible. It is quite apparent, however, that in these cases there must be competition between normal substrate and imposter. Once the imposter molecules become incorporated into the nucleic acids the toxic molecules may or may not be replaceable.

2. *Enzyme Stimulation.* Not only may toxicants act by inhibiting enzymes, but they may also act by overstimulating enzymes. Such overstimulation may throw the normal cellular physiology completely out of balance. This abnormal unbalance may be analogous to an overdose of insulin in humans. Byrde et. al. (1956), with *in vitro* studies of enzymes from *Monilinia laxa* showed that copper greatly stimulates the activity of DPN oxidase, while *o*-phenylphenol greatly stimulates cytochrome c oxidase. Mahler et. al. (1955) suggest a plan to explain the burst of molecular oxygen released from systems containing metalloflavoprotein enzymes (EF-Fe) when they are acted upon by quinones. Normally the electron transfer goes DPNH $\rightarrow$ EF–Fe $\rightarrow$ Cytochrome c. With quinone in the system, the electrons may be diverted completely from EF-Fe to molecular oxygen rather than cytochrome c, thus uncoupling all subsequent phosphorylations.

C. Chelation

Before presenting chelation as a toxic mechanism it would be well to know something about chelates.

Chelates are cyclic structures formed by the union of metallic atoms with inorganic or organic molecules or ions. The term comes from *chela*, the Greek name for crab's claw. One molecule of the nonmetallic portion of the complex is called a ligand. Compounds which chelate must be able to combine at more than one point with metals. A simple 2 point, or bidentate, ligand is pictured in Fig. 8.4. Each 8-quinolinol molecule is a ligand.

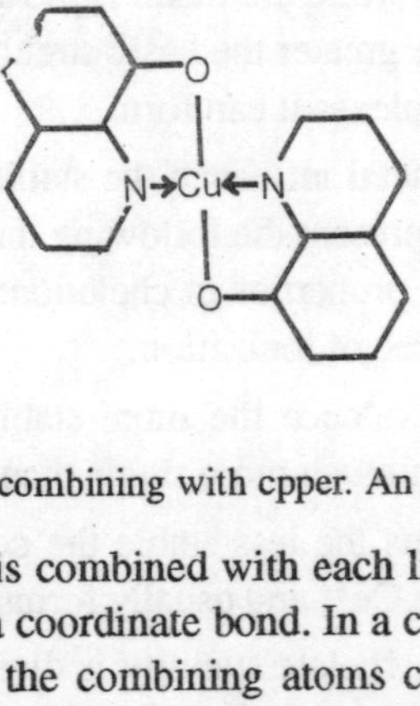

Fig. 8.4. 8-Quinolinol combining with cpper. An example of bidentate ligands.

Note that the metal is combined with each ligand by two bonds: (*a*) an ionic bond, and (*b*) a coordinate bond. In a coordinate bond, both of the electrons shared by the combining atoms come from only one of the atoms, nitrogen in this case. Other chelates are known in which all the bonds between ligands and metal are coordinate, without any ionic bonding.

It is the multiple bonding feature of chelation which gives chelators their most interesting property. This property is the formation of an extremely stable complex which can hold metals so tightly that they are released or exchanged only with great difficulty.

The formation of chelate compounds is usually accompanied by color formation and the loss of polar properties by the ligand molecules. As the atoms on the ligand which combine with metals are usually nitrogen, oxygen, or sulfur, it is quite apparent why polar properties are lost. Whether or not the complex remains water-soluble depends on whether or not the ligand has polar groups not involved in chelation. For example, copper 8-quinolinolate is quite insoluble in water, while copper 8-quinolinol-5-sulfonate is water-soluble.

The smaller chelated cyclic structures are considered to be coplanar. According to the Baeyer strain theory, these, like other organic cyclic compounds, would be most stable when there are 5 or 6 atoms in each ring. Chelate cycles with less than 4 atoms or more than 7 atoms are known, but they are uncommon and usually unstable. Because of certain peculiarities of chelate rings, 4-membered structures are more common and more stable than 4-membered carbon rings. The probability of a compound being a chelator is highest if it has 2-electron-sharing atoms separated by 2 or 3 other atoms.

In the union of ligand with metal, the ligand is considered to be the electron-pair donor, or base, while the metal is the acceptor, or acid. All other things being equal, the greater the basic strength of the ligand, the more stable the chelate complexes it can form.

What properties of the metal influence the stability of the complex? Irving and Williams (1953) present the following three features as being the most important metallic properties in chelation: (*a*) valence, (*b*) the ionic radius, and (*c*) the degree of ionization.

The higher the cationic valence the more stable the complex. For example, F^{+++} complexes are much more stable than Fe^{++} complexes.

The larger the ionic radius the less stable the complex. Mn^{++} has a much larger ionic radius than Cu^{++} and usually forms less stable chelates.

Other things being equal, chelate stability is directly correlated with the second ionization potentials of the metals.

Chelating compounds are both common and important in biology. Among them are organic acids, amino acids, enzymes, vitamins, hemoglobin, and chlorophyll.

Chelation as a toxic mechanism has been a subject of active research and debate since it was first proposed by Zentmyer for fungi (1943) and by Albert for bacteria (1944).

Zentmyer (1944) showed that the toxicity of the powerful chelator 8-quinolinol (oxine) to the fungus can be overcome by the addition of excess zinc to the growth medium. Furthermore, reducing the pH of the medium to an acidity at which oxine cannot chelate, also destroyed its fungitoxicity. On this evidence, Zentmyer related the fungitoxicity of oxine to its ability to chelate, and proposed that oxine poisons by robbing the organism of required trace metals. Albert and his group were also able to destroy the antibacterial action of oxine by the addition of excess metal, iron in this case. In addition, they found that isomers or analogues oxine able to chelate, were also antibacterial. Those related compounds unable to chelate, e.g,, 2-quinolinol, were not antibacterial.

Soon, however, the metal-robbing theory was seriously questioned by Mason (1948) who showed that copper oxinate is more fungitoxic than the unchelated oxine. This was particularly damaging to the theory of metal-robbing. Since the copper chelate is one of the most stable of oxinates, it would not be expected to chelate other trace metals. Albert and his group found that oxine is not antibacterial unless iron is present in trace amounts. It must be made clear here that chelation as a toxic mechanism was not questioned. In doubt was only the theory that the toxicity of oxine derives from metal-robbing. It is ironic that the metal-robbing theory of toxicity should have been based on oxine which may or may not work in that way. There is not much doubt that chelators may exert their toxicity by metal-robbing. For example, Rich and Horsfall (1948) have shown that dimethylglyoxime will produce trace metal deficiency symptoms in *Aspergillus niger*. On dimethylglyoxime, the fungus grows in curds, rather than in a normal mat, and sporulation is poor or absent.

In what other way may chelators or chelate compounds poison? Albert and others have suggested that the poisonous form of oxine is the half chelate. Goksøyr (1955) also proposes the half chelate as the toxic species of dithiocarbamates. If this is so, then the half chelates may be poisoning by blocking functional groups. For example, an amino group and hydroxyl group properly spaced on the surface of an enzyme protein would form a mixed chelate with the half chelated toxicants. This is illustrated in Fig. 8.5.

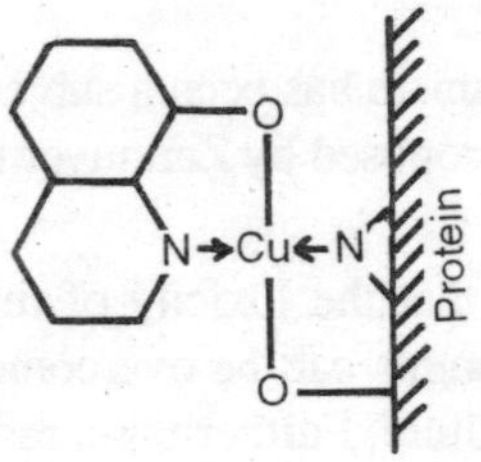

Fig. 8.5. Half chelate blocking functional groups.

The same effect could be had with a sulfhydryl group in place of the amino or hydroxyl group.

Another type of functional group blocking would be the formation of mixed chelates with the phosphoric acid or nitrogen base portions of nucleotides. This may be illustrated (Fig. 8.6) with copper oxinate and adenosine-5′-diphosphate (ADP).

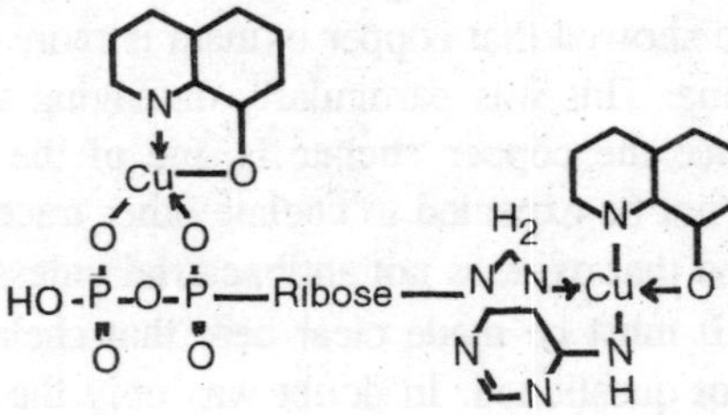

Fig. 8.6. Half chelate blocking ADP.

So much for chelator and their possible toxic mechanisms. The fascinating stories of oxine and dithiocarbamates as toxicants will be described in detail in a later section.

D. Selective Solubility

Toxic compounds may act by selective solubility. Ferguson (1939) has shown that the insecticidal value of man chemically "inert" materials, such as saturated hydrocarbons, is correlated with their relative solubility in the medium in which they are applied. He pointed out that the partition of these compounds between the ambient medium and the biophase will be proportional to the degree of saturation of the ambinet medium. In other words, the less soluble the toxicant in the applying medium, the greater the saturation of that medium and the greater the proportion entering the biophase. The degree of toxicity of these structurally non-specific compounds is directly proportional to their saturation of the medium in which they are applied. For example, nonspecific compounds X, Y, and Z may require respectively 1000 p.p.m., 100 p.p.m, and 1 p.p.m. to half-saturate the medium in which they are applied, whether it be water or air. These dosages should approximate half the maximum toxicity possible for treating organisms with compounds X, Y, and Z.

Ferguson was able to use this effect to explain another situation common in toxicity studies. Very often in comparing homologues of a toxic molecule, toxicity increases as the alkyl side chain lengthens. This continues until the addition of one more methyl group makes that homologue almost nontoxic. Ferguson suggests that the final methyl group would make the compound so insoluble in the applying fluid, that the amount required for complete saturation would partition less than a toxic dose into the biophase. This cut-off point will vary with compounds, applying medium, and tissue poisoned.

Ross and Ludwig (1957) have presented evidence that the fungitoxicity to the *N-n*-alkylethylenethioureas follows the Ferguson effect.

All this is interesting and useful, but dose not explain how supposedly "inert, non-specific" compounds can poison. There are at least two possibilities.

The first is a solvent effect. If the inert molecules partially replace and force apart the biophase molecules, metabolic reaction rates could be altered enormously. Taking a test-tube example from organic chemistry, the reaction of ethyl iodide with triethylamine is only 1/1000 times as fast in hexane as it is in benzyl alcohol. The nonspecific toxicants could slow down vital reaction rates by reducing the frequency with which substrate molecules must collide with enzyme molecules essential for continuing normal processes.

The second way in which nonspecific toxicants may act is by altering the colloidal structure of the semipermeable of the cell. Many of these toxicants have a particular avidity for the lipoproteins of the cellular membranes. They are also known to exert cellular narcosis, allowing cellular contents to leak from the cell. The most likely way that these compounds could do this is by affecting the structure of the hydrophilic colloids making up the semipermeable membranes.

E. Dissolution

Toxicants may act by dissolving cellular components necessary to the physical structure of the cell. A good example is lysozyme, an antibiotic enzyme discovered by Fleming (1922) in mucous secretions. Lysozyme is a mucopolysaccharide depolymerase. It is capable of completely dissolving the cell walls of bacteria. Commercial lysozyme, prepared from egg whites, is assayed by its ability to dissolve suspensions of the bacterium *Micrococcus lysodeikticus*. The end-product of lysozyme activity is *N*-acetylglucosamine. There is no increase in reducing groups. Lysozyme has been found in many bacteria, including those susceptible to its activity. It also occurs in *Ficus latex* and in commercial papain.

Rich has found that lysozyme, with no apparent effect on spore walls, will lyse the germ tubes of *Stemphylium sarcinaeforme*, allowing the cellular contents to escape. Rich also found that sodium hypochlorite causes the separation of the cells of this multicelled spore.

Carter and Lockwood (1957) described the lysis of the mycelium of *Glomerella cingulata* by a diffusate from *Streptomyces* sp. They also report the lysis of *G. cingulata* mycelium by antibiotics such as mycostatin and thiolutin, and by fungicides such as ferbam, thiram, and zineb. The latter effect that ascribe to a hastening of autolysis of the poisoned mycelium, rather than a direct dissolution of the mycelium by the fungicide.

A material may not cause a complete dissolution of cell wall, but may instead have a partial effect on its rigidity. Griseofulvin may act in this fashion. This antibiotic was found to be the "curling factor" produced by *Penicillium janczewskii*. It causes a spiral growth of the germ tubes of *Botrytis allii* and other fungi. This abnormality of the germ tubes apparently prevents the organism from developing mycelium.

F. Precipitation

Precipitation, the opposite of dissolution, may poison either by making metabolites unavailable or by destroying colloidal structure.

Polysulfides, present in lime sulfur, may produce their poisonous effect by precipitating essential metals and so making them unavailable. *Aspergillus niger*, grown in the presence of potassium sulfide, will produce yellow, rather than black, spores. This is a symptom of metal deficiency. Hydroxides or carbonates may also produce this effect.

Precipitation or coagulation of hydrophobic colloids, such as proteins, would also have a profound effect on normal cellular processes. The ability of a compound to bring about coagulation will depend on two things. First, the power to neutralize the charge on the dispersed phase, or the lowering of the zeta potential; and, second, the ability of the coagulant to take the stabilizing hydration water from the dispersed particles.

The ability to neutralize charges will depend on the valency of the ion charge oppositely to that of the dispersed phase. For example, if a protein sol is at a pH on the alkaline side of its isoelectric point, the dispersed phase will be negatively charged. The ability of the coagulant to reduce the charge on the protein micelle will be directly related to the valency of the cation of the coagulant. Actually, very small amounts of oppositely charged ions may cause a large drop in the zeta potential. The hydrophilic colloid may increase in viscosity as its charge is neutralized, but coagulation is prevented by the stabilizing effect of the hydrated shells around each micelle.

The final phase of coagulation depends on the ability of the toxicant to dehydrate. This ability to take water from the particles usually follows the lyotropic series. For the effect of anions to precipitate proteins, this series is usually shown as follows: citrate > tartrate > SO_4 > acetate > Cl > NO_3 > Br > I > CNS. Citrate shows the greatest effect and thiocyanate the least. This series for cations is, Th > Al > H > Sr > Ca > K > Na > Li. Ethanol and formaldehyde are two other toxicants which may act as protein coagulants.

G. Hydrolysis

Acids or bases may poison by hydrolysis. Hydrolysis is the breaking of the reacting molecule and the combination of the pieces with H^+ and OH^- from water.

From the viewpoint of toxicity, proteins are probably the most important biological compounds which may be hydrolyzed. Hydrolysis would, of course, be a denaturing process, particularly disruptive to enzymes. When enzymes are hydrolyzed, the most profound effect is the breaking of peptide linkages which join amino acids to make the peptide chains of which proteins are built (Equation 4).

Mandels (1953) used weak acid treatment to destroy the activity of enzymes presumably located at or near the surface of *Myrothecium* spores.

$$H_2N{-}R{-}\overset{\displaystyle O}{\overset{\|}{C}}{-}\underset{\displaystyle H}{N}{-}R'{-}\overset{\displaystyle O}{\overset{\|}{C}}{-}OH + H_2O \xrightarrow[\text{base}]{\text{acid or}}$$

$$H_2N{-}R{-}\overset{\displaystyle O}{\overset{\|}{C}}{-}OH + H_2N{-}R'{-}\overset{\displaystyle O}{\overset{\|}{C}}{-}OH \qquad \text{(Eq. 4)}$$

Blood (1937) suggested that the acid production by fermenting tomato pulp is responsible for the elimination of *Corynebacterium michiganense* from tomato seed. He showed that soaking infested seed in acetic or lactic acid does as well as the fermentation process.

Tyner (1951) made the surprising discovery that merely soaking barley seed in water for a period of time can control loose smut. An investigation of this phenomenon by Leben et. al. (1956), and Jacquent et. al. (1957) showed that the control is achieved by organic acids. These organic acids, acetic, butyric, and formic, produced by bacteria operating during the water soaking, are the actual toxicants.

H. Oxidation-Reduction

Oxidation-reduction changes, in order to cause any real damage to cells, must be drastic rather than mild. Living cells usually have too large a capacity to equilibrate changes in oxidation-reduction to expect any lasting toxicity from mild changes. Calcium hypochlorite, a strong oxidant, is very toxic. Whether or not it acts on a specific system is not important. It is probably nonspecific by virtue of its capacity to oxidize strongly many vital components within the cell. Ozone, another strong oxidant, is toxic to many types of cells. Its production by ultraviolet radiation may account in part for the deleterious effects of UV light. Such systems as Cu^{++} Cu^{+} may be poisonous by virtue of their oxidant action. For example, Albert (1952) reported that Cu^{++} reacts rapidly with cysteine to give cystine and cuprous cysteinate. Other toxic oxidizing agents are SO_2, H_2O_2, potassium permanganate, sodium perborate, and sodium dichromate. Elemental sulfur may poison by acting as an oxidant.

Horsfal (1945) suggested that the dinitrophenols are toxic because they can oxidize and so split unsaturated fatty acids. Dinitrophenols are very effective against powdery mildews. Horsfall later (1956) proposed that the compounds kill conidia of powdery mildew by destroying the fatty envelope that retains the water these spores need to survive. The spores then die by desiccation.

Hydrogen sulfide is a powerful reducing agent particularly toxic to cells. It may act not only by precipitating metals, but also by holding them in a reduced state.

Oxidation-reduction systems may show definite antibacterial action not shown by either component of the system. Guest and Salle (1942) found that a combination of Fe Cl_2 with either Fe Cl_3 or $Fe_2 (SO_4)_3$ is much more antibacterial than any of these compounds used separately. This was also true of Sn Cl_2 with Sn Cl_4, Fe Cl_2 with Sn Cl_4, Sn Cl_2 with Fe Cl_3, Mn SO_4 with Sn Cl_4, and Mn SO_4 with $Fe_2 (SO_4)_3$. These heavy metal salts were most toxic when mixed in certain proportions. The increase in toxicity required that the mixture had one metal at a higher oxidation state than the other. Only the salts of heavy metals gave this effect.

I. Surface Activity

Surface activity is a property of any substance whose presence in small amounts greatly alters the surface behavior of a given system. "Greatly alters" is rather a loose term, as most solutes will alter the surface behavior of the solvent to some extent. Although there are water-insoluble surfactants, we will be concerned with water-soluble organic compounds that markedly lower the surface tension of the liquid system in which they are dissolved. The most common of these surfactants is, of course, soap. A burgeoning chemical industry has produced a great number of synthetic surfactants that have found uses in all phases of our daily life.

What are the molecular properties that make a surfactant? Basically, a surfactant molecule is a long molecule, one end polar and the other end nonpolar. The polar end has an affinity for polar substances, such as water, while the nonpolar end has an affinity for nonpolar substances, such as grease, fats, and oils. When surfactant-containing water is applied to grease particles, the nonpolar end of the surfactant molecules is adsorbed by the grease. Since the polar end will not dissolve in the grease, the surfactant molecules stay at the surface. The surface-located polar ends having an affinity for water give the grease particle a miscibility with water. This allows the grease particles to become suspended in water, or emulsified, and so removed.

Surfactants lower the surface tension of water because the nonpolar ends of the surfactant molecules are forced out of water, and are concentrated and arranged linearly at right angles to the surface of the liquid, the polar end toward the liquid and the nonpolar end forced away from the liquid. The surface then acquires primarily the nonpolar properties of he surfactant. As the attraction between surfactant molecules is much less than the attraction of one water molecule for another water molecule, the surface tension is lowered. The reason why only a small amount of surfactant is required to that it is forced to the liquid-air or liquid-solid interface and all that is required for maximum effect is a unimolecular layer of the surfactant molecules at the interface.

There are three kinds of surfactants, anionic, cationic, and nonionic. The ionic surfactants ionize or dissociate into two parts, a large fatty radical and a small salt radical. If the change in the fatty radical is negative, the compound is anionic. For example, a soap dissociates into $C_{17}H_{35}COO^-$ and Na^+, typical of an anionic surfactant. If the charge on the large radical is positive, the compound is classed as cationic. Cetylpyridinium chloride ionizes into

$$\left[H_{33}C_{16}\text{-}N\bigcirc\right]^+ \text{ and } Cl^-,$$

making it a cationic surfactant. Nonionic, or ampholytic surfactants from zwitterions in solution. An example is cetylaminoacetic acid: $H_{33}C_{16}NH^+—COO^-$.

Soap is a weak germicide. Its primary value as a sanitizer is to remove germs physically rather than to kill them. The real interest in the toxic and germicidal effects of surfactants dates from the work of Domagk (1935) who found that some cationic surfactants are powerful antibacterial compounds. Since then a number of quaternary ammonium compounds have been developed both as bactericides and fungicides. These have been principally alkylpyridinium halides and alkyl trimethylammonium halides. These cationic compounds are effective against both gram-positive and gram-netative bacteria and against fungi. Schwartz et. al. (1958) list 43 different types of cationic surfactant germicides.

Anionic surfactants are primarily active against gram-positive bacteria. Examples of anionic surfactant toxicants are soaps and certain aliphatic carboxylic acids.

Nonionic surfactants usually have little or no germicidal action. Schwartz et. al. (1958) list the salt of dioctylaminoethylglycine as examples of nonionic germicidal surfactants.

Surfactants appear to exert their toxic effects by breaking cross-links between polypeptide chains, or by disrupting colloidal structures. The effect on polypeptide structures is evident in the report of Sreenivasaya and Pirie (1938) who found that the protein of tobacco mosaic virus is disintegrated by surfactants. Anson (1939) also found that surfactants could split proteins into much smaller molecular units.

The profound effect of surfactants on colloidal structure is noted in toxicity studies by a disruption of the semipermeable membranes allowing the leaking materials from the cell. Horsfall (1956) discusses the importance of this as a toxic mechanism in both bacteria and fungi. Salton et. al. (1951) have taken electron micrographs of bacteria treated with cetyltrimethylammonium bromide and have observed morphological changes correlated with cell death. Bacteria treated with 45 mg of cetyltrimethylammonium bromide per 1.5 mg. dry weight of bacteria for 5 minutes show a pulling away of the cytoplasmic membrane from the cell wall. This dosage of cetyltrimethylammonium bromide just initiates leaking. The proportion of cells (30%) killed by this treatment is the same as the number of treated cells showing contracted cytoplasm. Treating the cells with 900 mg. cetyltrimethylammonium bromide per 1.5 mg. of bacteria for 30 minutes strips off the cell walls.

J. Osmotic Pressure Changes

Probably the oldest means of killing microorganisms is the preservation of foods with compounds that produce high osmotic pressures. Salting and sugaring are ancient means of preventing the growth of molds and bacteria on foods. Although the effective dosages of these materials are very high, salt and sugar were usually close at hand. On a dosage basis, this toxic mechanism is very inefficient, and will probably not be of importance in the development of future fungicides.

There are probably other toxic mechanisms of importance, but those listed there undoubtedly include those of principal interest.

It is entirely possible for a single toxicant to have more than one mode of action, one of which will be effective on one organism and another mode on a second organism. One can picture highly reactive molecules, such as captan, passing into the cell and reacting with metabolites *en route*. Toxicity or death of the cell results when any vital link is broken.

II. INORGANIC FUNGTOXICANTS

A. Metals

The most widely used inorganic toxicants are metallic compounds. Of the metals used as fungitoxicants, copper, specifically Bordeaux mixture, has been used in the largest amounts. Actually, in terms of minimum toxic dosages, inorganic metallic compounds are not nearly so toxic as a multitude of organic, nonmetallic toxicants. Even the organic metal complexes are more potent than the inorganic compounds containing the same metals.

According to Horsfall (1956), the descending order of fungitoxicity for the metal cations is approximately Ag > Hg > Cu > Cd > Cr > Ni > Pb > Co > Zn > Fe > Ca. McCallan and Wilcoxon (1934) found also that osmium and ruthenium are quite fungitoxic. Aluminum and palladium are only slightly toxic, and rhodium and iridium are essentially non-fungitoxic. McCallan and Wilcoxon (1934) studies the fungitoxicity of the elements in relation to their position on the periodic table. They generalized that toxicity with a group usually increased with increasing atomic weight. The principal exceptions are silver, much more toxic than gold, and chromium, usually more toxic than molybdenum and tungsten. A study of the periodic table shows that most toxic metals fall in groups I, II, and VIII.

Horsfall (1956) pointed out the good correlation between the toxicity of metals and their ability to form stable chelate compounds. His comparison of chelate stability with the toxicity of the metals is given below.

Toxicity: Hg > Cu > Cd > Ni > Pb > Co > Zn > Fe > Ca.

Chelate stability: Hg > Cu > Ni > Pb > Co = Zn > Cd > Fe > Mg.

Horsfall regards this as evidence that the primary toxic mechanisms of metals is their ability to combine with metabolites, enzymes, and proteins. Palladium and silver are two glaring exceptions to Horsfall's generalization. Palladium, only weakly toxic, forms very stable chelates, while silver, highly toxic, forms only weak chelates.

Shaw (1954) correlated toxicity with the insolubility of the metal sulfide. His diagrams show excellent correlations between the negative logarithm of lethal dose and the solubility product constants of the various metallic sulfides. This correlation was fairly consistent for the poisoning of diastase, urease, the paramecium, a planarium (*Polycelis nigra*), or a fish (the stickeback). Shaw believes this is the result of the "thermodynamic affinity of the inhibitor [metallic cations] for key functional group in the enzyme's catalytically active site [sulfhydryl groups]."

Shaw, of course, was preceded by Fildes (1940) who showed that mercury toxicity could be alleviated by sulfhydryl-containing compounds. Fildes suggested that mercury at low concentration acts by inactivating sulfhydrul groups without other permanent injury to the cell. British anti-Lewisite (2, 4-dimercaptopropanol) was developed specifically as an antidote for the arsenical war-gas Lewisite Cl CH = CH.As Cl_2). The development of BAL was based on the discovery that arsenic poisons by knocking out sulfhydryl groups.

Sulfhydryl compounds will also reverse copper toxicity of peas and of fungus spores. Foster showed that three kinds of seed susceptible to copper injury contain sulfhydryl groups detectable with nitroprusside. Four kinds of seeds tolerant to copper contain no detectable sulfhydryl groups. Copper seed-treatments eliminated the detectable sulfydryl groups of the copper-sensitive seeds. Both Yoder and Foster concluded that the sulfhydryl groups poisoned were essential to the functioning of the respiratory enzymes.

Metallic compounds are often used as astringents in medicine. Astringency is the precipitation of proteins by high concentrations of metallic compounds.

An obvious question is the relation of toxicity to the oxidation state of the metal. Apparently no generalization can be made concerning this point. Cupric ions are fungitoxic, but cuprous oxide is more toxic than cupric oxide. Mercuric ions are usually more toxic than mercurous ions.

No discussion of the toxicity of cations would be complete without including something about sorption. Are cations adsorbed at the periphery of the cell or are they absorbed freely into the interior? Bodnar and Terenyi (1930) found that the uptake of copper by *Tilletia tritici* followed the Langmuir adsorption isotherm. Sussman and Lowry (1955) have strong evidence tat dormant ascospores of *Neurospora* also adsorb cations. McCallan (1948), working with copper sulfate on spores of *Alternaria oleraceae* and *Monilinia fructicola*, could not fit his data to Langmuir's isotherm. Miller et. al. (1953) do not believe that silver and cerium are merely adsorbed. As evidence, they cite the enormous quantities of cations taken up by their fungus spores. The number of cationic molecules taken up, if merely adsorbed, would form layers several molecules thick on the spore surface. In addition, a pretreatment with silver does not interfere with the uptake of cerium and vice versa. Interference would be expected if uptake of these cations were strictly by adsorption. The work of Owens and Miller (1957) also points to penetration of cations into spores with practically no concentration on the cell walls. Rothstein and Hayes (1956), however, find that yeast cells bind both univalent and bivalent cations at the cell periphery. The binding is rapid and reversible, with the bivalent ions being held more firmly than the univalent ions. They tentatively identify the binding sites as phosphoryl and carboxyl groups. These sites are apparently separated from the interior of the cell by a permeability barrier. Rothstein and Hayes suggest that certain cations will have special affinites for specific binding groups on the cell surface: UO_2^{++} for carboxyl groups, Hg^{++} for sulfhydryl groups, and Zn^{++} for imidazole groups.

The evidence, then, is contradictory, but can be resolved. In organisms where there is a permeability barrier, adsorption is readily demonstrated. Where no permeability barrier exists or is destroyed by the toxicant, adsorption may play a role but is obscured by the free passage of cations into and out of the cell.

Another factor which confounds the sorption picture is the effect of metabolites leaking from the cell. Miller and McCallan (1957) have shown that treating spores of *M. fructicola* with toxic doses of Ag^+ for as little as 1.2 minutes increases their outward loss of phosphorous compounds. The loss of phosphorous compounds from these silver treated cells is 20 times that of the untreated spores. This mass of organic material capable of complexing will undoubtedly have a profound influence on the final location of cations on or in the spore.

B. Nonmetals

1. *Sulfur*. Of the nonmetals, elemental sulfur is the most important fungitoxicant. In spite, of its use as a pesticide for centuries, its method of toxic action is still a question for debate. Pollacci (1975) suggested that H_2S produced from sulfur is actually responsible for the fungitoxic action of sulfur. Young (1922) proposed pentathionic acid as the toxic product from sulfur. Pentathionic acid results from the interaction of H_2S and SO_2. The importance of pentathionic acid, SO_2, and SO_3 was largely discounted by Roach and Glynne (1928) and Wilcoxon and McCallan (1930). These workers found that the toxicity of these compounds is no more than could be accounted for by their hydrogen ion concentrations.

McCallan and Wilcoxon (1931) found that sulfur-sensitive spores produced large amounts of H_2S from sulfur. They reverted to Pollacci's idea that H_2, which they found to be enzymatically produced, was the cause of sulfure toxicity. When they again tested the H_2S theory, Miller et. al. (1953) found that colloidal sulfur is as much as 50 times as toxic as H_2S to sulfur-sensitive fungi. H_2S, they now believe, cannot be the toxic agent responsible for the toxicity of sulfur. They concluded that sulfur poisons because it is an active hydrogen acceptor and so interferes in normal hydrogenation reactions of the cell. The intensity of H_2S production is a measure of the intensity of the hydrogenation process within the cell.

The theory of sulfur poisoning by competing for hydrogen seems plausible, but is still open to question. For example, methylene blue and ascorbic acid, both excellent hydrogen acceptors, are only weakly fungitoxic. Miller et. al. (1953) use as supporting evidence the work of Sciarini and Nord (1943) who concluded that, in *Fusaria*, elemental sulfur competes for hydrogen freed by dehydrogenases. Although Sciarini and Nord present such a conclusion, it is not supported by their data.

All of these ideas are possible. What Miller et. al. (1953) have actually shown is that H_2S produced within the spore is more toxic than H_2S in the ambient fluid. This is not sufficient evidence to discard H_2S as the source of sulfur toxicity. H_2S which ionizes primarily as HS^- may well poison metal-activated compounds just as metals poison sulfhydryl-activated compounds.

Miller et. al. (1953) also discount H_2S as a factor because they find the H_2S given off pretty well accounts for all the sulfur added to the system. However, the amount of H_2S bounds *in situ* by the micro-quantities of metals in the spore would be so small that it would be practically undetectable.

McCallan and Miller (1956) made the significant discovery that fungi in reducing sulfur release one mole of extra CO_2 for each mole of H_2S produced. This contradicts Horsfall's (1956) suggestion that sulfur replaces oxygen in Krebs cycle intermediates. If this were so, then there should be less CO_2 than normal, rather than more. The production of CO_2 also indicates that elemental sulfur is not just competing for hydrogen, but appears to be acting as a strong oxidizing agent in the process of being reduced.

2. *Halogens*. Halogenated inorganic compounds are another group of useful toxicants. The commonest inorganic halogen compounds used as germicides are calcium or sodium hypochlorite. The toxicity of these compounds is directly correlated to the oxidizing power of their solutions. For example, Rideal and Evans (1921) showed that alkalinity which reduces the germicidal effect of hypochlorites also reduces their oxidizing power. The strength of the hypochlorites is expressed as "available chlorine", a measure of oxidizing power.

CaOCl reacts with water and CO_2 from the air as in Equation 5.

$$CaOCl + HCO_3^- \rightarrow CaHCO_3 + ClO^-$$
$$ClO^- \rightarrow Cl^- + O \uparrow \quad \text{(Eq. 5)}$$

Chlorine evolved from acidification of CaOCl, will also oxidize as in Equation 6.

$$Cl_2 + H_2O \rightarrow HCl + HOCl$$
$$HOCl \rightarrow HCl + O \uparrow \quad \text{(Eq. 6)}$$

This is the method by which chlorine gas purifies water. Halogens may also act by halogenating metabolites, thus producing antimetabolites *in situ*.

3. *Formaldehyde*. Formaldehyde (HCHO) is widely used as a soil fumigant. Its action is primarily as a protein denaturant. The combination of formaldehyde with amino acids, as in the classic Sörenson formol titration (Equation 7), is representative of the way formaldehyde may combine with and denature proteins.

$$\underset{\text{(Zwitterion)}}{{}^+NH_3CH(R)COO^-} + HCHO \rightarrow CH_2O.NH_2CH(R)COO^- + H^+ \quad \text{(Eq. 7)}$$

4. *Ammonia Gas*. Ammonia fumigation has recently been used successfully in citrus storages for reducing fruit rots. Ammonia, highly soluble in water with which it forms NH4OH, probably poisons fungi by drastic changes in their internal pH^-.

III. ORGANIC FUNGITOXICANTS

A. Dithiocarbamates

The greatest number of useful organic fungitoxicants has come from the dithiocarbamates. These compounds all have in common the grouping.

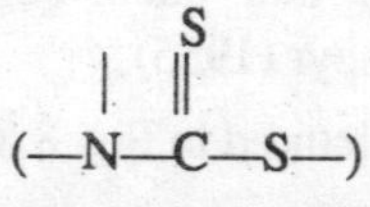

$$(-\overset{|}{N}-\overset{\overset{S}{\|}}{C}-S-)$$

One series, based on the patent issued to Tisdale and Williams (1934), is made from the alkaline interaction of CS_2 and simple amines. These compounds are the dialkyldithiocarbamates and thiuram sulfides. The second series, based on Hester's patent (1943), is made from the alkaline interaction of CS_2 and ethylenediamine. These are the ethylene bisdithiocarbamates.

1. *Dialkyldithiocarbamates and Thiuram Sulfides*. The most widely used of these materials are zinc dimethyldithiocarbamate (ziram), ferric dimethyldithiocarbamate (ferbam), and tetramethylthiuram disulfide (thiram). Recently, another member of the family, sodium *N*-methyldithicarbamate, has been finding use as a soil fungicide. The structure of these compounds are given in Fig. 8.7.

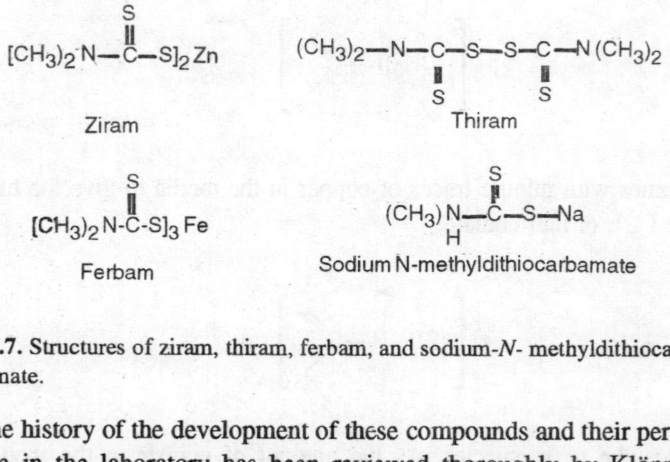

$$[(CH_3)_2N-\overset{\overset{S}{\|}}{C}-S]_2Zn$$

Ziram

$$(CH_3)_2-N-\underset{\underset{S}{\|}}{C}-S-S-\underset{\underset{S}{\|}}{C}-N(CH_3)_2$$

Thiram

$$[(CH_3)_2N\text{-}\overset{\overset{S}{\|}}{C}\text{-}S]_3Fe$$

Ferbam

$$(CH_3)\underset{H}{N}-\overset{\overset{S}{\|}}{C}-S-Na$$

Sodium N-methyldithiocarbamate

Fig. 8.7. Structures of ziram, thiram, ferbam, and sodium-*N*- methyldithiocarbamate.

The history of the development of these compounds and their performance in the laboratory has been reviewed thoroughly by Klöpping (1951), Horsfall (1956), and McCallan (1957).

The two most important clues to the toxic action of the dialkyl dithiocarbamates and the thiuram disulfides are (*a*) their formation of metal chelates and (*b*) their polymodal dosage response. The first property is a well-known one, commonly used in analytical chemistry. The second was first reported by Dimond et. al. (1941), for thiram. This latter effect, also called "inversion" and "quenching", is described in detail and logically explained by Goksøyr (1955).

Inversion of toxicity is pictured in Fig. 8.8.

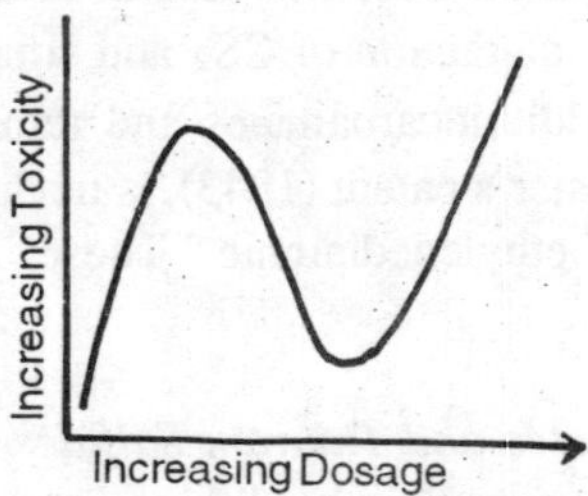

Fig. 8.8. Polymodal dosage response curve, or inversion of toxicity.

The zone of inversion, where increasing dosage actually depresses toxicity, has been particularly hard to explain. Goksøyr's explanation is as follows. The dimethyldithiocarbamate ion,

$$\left[(CH_3)N{-}C\begin{matrix}{\nearrow\!\!\!\!/ S}\\{\searrow S}\end{matrix}\right]^-$$

combines with minute traces of copper in the media to give the highly toxic 1 : 1, or half chelate,

$$\left[(CH_3)N{-}C\begin{matrix}{\nearrow\!\!\!\!/ S}\\{\searrow S}\end{matrix}\right]Cu^+$$

This is the toxic species. As the amount of copper in the media is constant, the addition of more dithiocarbamate forms the less toxic 1 : 2, or full chelate,

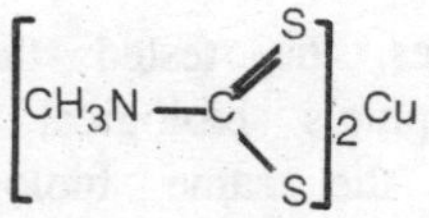

$$\left[CH_3N{-}C\begin{matrix}\nearrow S\\ \searrow S\end{matrix}\right]_2Cu$$

Toxicity falls. As more dithiocarbamate is added, the intrinsic toxicity of the dimethyldithiocarbamyl ion exerts itself, acting perhaps by binding essential metallic micronutrients. Thus toxicity increases again.

Goksøyr demonstrated that yeast cells reduce thiram to dimethyldithiocarbamyl ions. Hence, thiram activity and its polymodal dosage response curve is identical to those of the dialkyldithiocarbamates.

As will be seen later, *N*-methyldithiocarbamate has an action more like the ethylenebisdithiocarbamates than like the dialkydithiocarbamates.

2. *Bisdithiocarbamates.* This group of compounds arose around disodium ethylenebis (dithiocarbamate), now called nabam (Fig. 8.9).

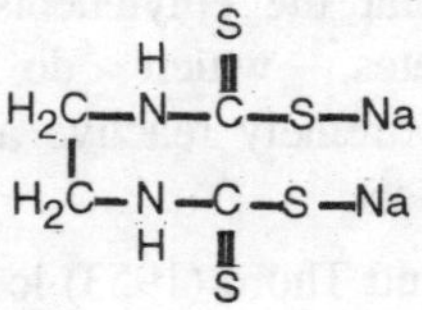

$$\begin{matrix} & H & S & & \\ & & \| & & \\ H_2C{-} & N{-} & C{-} & S{-} & Na \\ | & & & & \\ H_2C{-} & N{-} & C{-} & S{-} & Na \\ & H & \| & & \\ & & S & & \end{matrix}$$

Fig. 8.9. Nabam.

The history of their development and the laboratory results of structure-activity studies are given in Barratt and Horsfall (1947), and Klöpping (1951).

What is the basis for their activity? Barratt and Horsfall (1947) proposed that since these compounds readily release H_2S, they poison by means of the H_2S evolved. Rich and Horsfall (1950) demonstrated that H_2S could not account for the toxicity of nabam. The clue to the currently accepted mechanism was noted in the title of the paper by Dimond et. al. (1943), in the first report on nabam. They called it a "water-soluble fungicide with tenacity." This normaly puzzled them. In the data of Barratt and Horsfall appeared another clue. Zineb, the zinc salt of nabam, becomes more toxic upon aging. Klöpping (1951) blazed the trial which led to the present theory. Realizing that diisothiocyanates could form from bisdithiocarbamates, he tested the fungitoxicity of the diisothiocyanates. Klöpping's biological tests showed that the diisothiocyanates have the same biological spectrum as the bisdithiocarbamates. By "biological spectrum" he meant that different fungi have the same relative susceptibility to both types of compounds. In addition, the diisothiocyanates are much more toxic than the corresponding bisdithiocarbamates. Data from Klöpping (Table 8.1) are illustrative.

TABLE 8.1

Fungitoxicity Data from Klöpping Illustrating Similarity Between the "Biological Spectra" of Nabam and Ethylene Diisothiocyanate

	Minimum p.p.m. completely inhibiting growth			
Compound	*B. cinerea*	*P. ilalicum*	*A. niger*	*R. nigricans*
Nabam	0.1	0.1	0.5	20.0
Ethylenediisothiocyanate	0.05	0.02	0.05	10.0

Kløpping concluded that the ethylenebisdithiocarbamates act by releading diisothiocyanates, which do the poisoning. The diisothiocyanates being extremely reactive and unstable, are hard to detect as breakdown products.

Shortly after, Ludwig and Thorn (1953) looking for insoluble (tenacious) compounds from the oxidation of nabam, found that ethylenethiuram monosulfide is a major product of nabam oxidation. They found that ethylenethiuram monosulfide (Fig. 8.10) is highly fungitoxic.

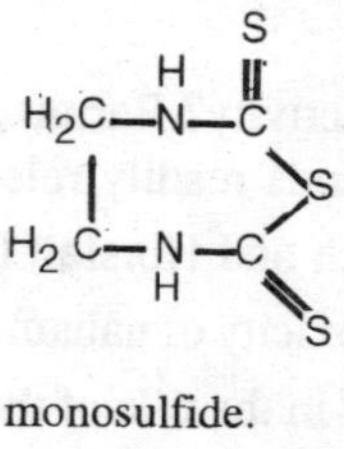

Fig. 8.10. Ethylenethiuram monosulfide.

Ethylenethiuram monosulfide they concluded to be the toxic breakdown product accounting for both the tenacity and toxicity of nabam. In a later paper, Ludwig et. al. (1955) reported that metallic ions catalyze the formation of isothiocyanates from ethylenethiuram nonosulfide. They could demonstrate this only in a nonaqueous medium (chloroform).

In return, Sijpesteijn and van der Kerk (1954) confirmed the production of ethylenethiuram monosulfide from nabam.

The present theory of nabam action may be summarized from the literature. Freshly prepared, unoxidized nabam is presumed to be essentially nontoxic.

The acidic breakdown of nabam proceeds as in Equation 8.

$$\begin{array}{l} H_2C-NH-C(=S)-S\,Na \\ | \\ H_2C-NH-C(=S)-S\,Na \end{array} \xrightarrow{\text{acid}} \begin{array}{l} H_2C-NH \\ | \qquad\quad \rangle C{=}S \\ H_2C-NH \end{array} + CS_2 + H_2S + 2Na^+$$

Nabam — Ethylenethiourea (nontoxic)

Basic hydrolysis and oxidation of nabam proceeds in another fashion (Eqation 9).

$$\begin{array}{l} H_2C-NH-C(=S)-SNa \\ | \\ H_2C-NH-C(=S)-SNa \end{array} \xrightarrow[\text{Oxid}]{\text{Base}} \begin{array}{l} H_2C-NH-C(=S) \\ | \qquad\qquad\quad \rangle S \\ H_2C-NH-C(=S) \end{array} + S + 2Na^+$$

Nabam — Ethylenethiuram (monosulfide monomer or polymer)

Ethylenethiuram monosulfide in turn breaks down to release ethylenediisothiocyanate (Eqation 10).

$$\begin{array}{l} H_2C-NH-C(=S) \\ | \qquad\qquad\quad \rangle S \\ H_2C-NH-C(=S) \end{array} \xrightarrow{\text{(metal)}} \begin{array}{l} H_2C-N{=}C{=}S \\ | \\ H_2C-N{=}C{=}S \end{array} + S$$

Ethylenethiuram monosulfide — Ethylenedisothiocyanate

The diisothiocyanates, highly reactive and unstable, apparently combine with sulfhydryl groups to act as nonspecific poisons.

One of the unresolved problems of nabam toxicity is the nature of the gaseous toxicant which it evolves. This unidentified gas was first detected by Rich and Horsfall (1950), who demonstrated that the gas could not be H_2S or SO_2. They did not consider CS_2 as a possibility because it is a relatively weak toxicant. In 1951, Cox et. al. showed that CS_2 and ethylenediamine evolve from nabam hydrolysis. They proposed that these two gases are the toxic vapors, recombining within the spore to give nabam. Lopatecki and Newton (1952) argued that the toxic vapors are simply CS_2 evolved from nabam. Later work on the problem by Weed et. al. (1953) practically ruled out CS_2 or ethylenediamine as factors. These workers also doubted the possibility of the toxic gases being a diisothiocyanate. Weed et. al. (1953) conclude that "an unidentified volatile constituent is involved" in the toxicity of nabam. There the problem stands.

Two other types of compounds must be included here although they are not ethylenebisdithiocarbamates. These are sodium *N*-methyldithiocarbamate, as in Equation 11. Sodium *N*-methyldithiocarba-

$$H_3C-\overset{H}{N}-\underset{\underset{S}{\|}}{C}-S^{(-)} \rightarrow HS^{(-)} + CH_3N = C = S$$

N-Methyldithiocarbamate ion   Methylisothiocyanate

mate is used as a soil fungicide under the trade names of "Vapam" or "VPM."

3, 5-Dimethyltetrahydro 1, 3-5, 2-thiadiazine-2-thione, the active ingredient of the soil fumigant "Mylone", also breaks down to give methylisothiocyanate as in Equation 12.

$$\text{(ring: } S,\ H_2C,\ C{=}S,\ H_3C-N,\ N-CH_3,\ CH_2\text{)} + 2H_2O \xrightarrow{\text{soil}}$$

Another group of related fungitoxicants, the rhodanines, was described by van der Kert et. al. (1953). The structure activity of the rhodanines was also reported in 1953 by Brown and her co-workers

(1953). Van der Kerk reports (1956) that the rhodanines also are isothiocyanate fungicides and act according to Equation 13.

$$\text{R—N—C=O, C(=S)—S—CH}_2 \text{ (rhodanine ring)} \underset{H^+}{\overset{OH^-}{\rightleftharpoons}} \text{R—NH—C(=S)—S—COO}^- \rightleftharpoons \text{R—N=C=S} + \text{HS—CH}_2\text{COO}^-$$

Although the dithioacarbamates have been the most useful and interesting of the organic sulfur fungicides, other also have been found to be fungitoxic.

Some of these are the ethylenethioureas, the xanthates, 2-mercaptobenzothiazole, the thiobisphenols, and those toxicants having a trichloromethylthio group.

Rich and Horsfall noted that the *N*-alkylethylenethioureas increased in fungitoxicity with increasing length of alkyl chain. Fungitoxicity is maximal at the 8 carbon homologue and decreased as the chain lengthens beyond 8 carbons. Ross and Ludwig (1957) demonstrated that this series borbably acts as physical toxicants exhibiting the Ferguson effect.

The xanthates, long used as disinfectants, were thoroughly investigated by Davies and Sexton (1946). They concluded that sulfur atoms on the C═S groups were not essential to the toxicity of xanthates. This conclusion was based on the essentially equal fungitoxicity of the following three compounds:

$$(C_2H_5O-\overset{S}{\overset{\|}{C}})_2S;\ (C_2H_5O-\overset{O}{\overset{\|}{C}})_2S \text{ and } C_2H_5O-\overset{S}{\overset{\|}{C}}-S-\overset{O}{\overset{\|}{C}}-OC-H_5$$

2-Mercaptobenzothiazole, although it is somewhat fungicidal, has found its principal use as a synergist with dialkyldithiocarbamates in proprietary mixtures sold as "Vancides."

The thiobisphenols will be discussed under phenolic fungicides. The trichloromethylthio compounds will be discussed under heterocyclic nitrogen compounds because of the interest in captan, the tetrahydrophthalimide derivative.

C. Heterocyclic Nitrogen Compounds

Horsfall and Rich (1951) reported on the structure-activity relationships in a large group of heterocyclic nitrogen compounds. Many of these compounds are extremely fungitoxic. Their toxic mechanisms, however, differ widely. Certain of them, such as the ethylenethioureas just discussed, the nitrosopyrazoles, and the tetrahydropyrimidines, probably act as physical toxicants. Others, such as the 2-imidazolines have been reported to be competitive inhibitors of enzymes. The action of acridine was discussed in an earlier section. Two other mechanisms may also be important to the fungitoxicity of heterocyclic nitrogen compounds. These latter two mechanisms may best be presented by discussing two compounds for which these mechanisms have been proposed. The compounds are 8-quinolinol (oxine), and *N*(trichloromethylthio)-4-cyclohexene-1, 2-dicarboximide (captan). Oxine has been introduced in the previous section on chelation, and will be discussed first.

Oxine is the cornerstone for the theories of toxic mechanisms involving chelation. As mentioned earlier, the ideal that oxine takes needed metals from organisms was called into doubt by evidence that the metal oxinates, particularly cupric oxinate, is more toxic than oxine itself. This evidence was further supported by Anderson and Swaby (1951), who found that oxine is a copper-deficient medium is only 1/48 as toxic as it is in a copper-sufficient medium. Copper must be present, therefore, for maximum fungitoxicity of oxine. Not only must there be copper present but the oxine must be able to form a chelate with it. For example, at a pH below which oxine chelates, it is no longer toxic, regardless of the presence of copper. The formation of a copper oxinate, then, appears to be essential to the fungitoxicity of oxine.

Complications arose. Block (1957) showed that the fungitoxicity of copper oxinate can be reversed not only with excess copper but also with excess oxine. Albert et. al. (1953) had reported earlier that the antibacterial action of oxine can be quenched by an excess of oxine—the polymodal dosage response again. They proposed that excess oxine inhibited the formation within the cell of the 1 : 1, or half chelate which they suggested as the toxic form of oxine. Reversal by excess metal, they argued, was caused by inhibiting the 1 : 2 or full chelate in the ambient fluid. They pictured the process as follows. Oxine, because of water solubility, penetrates into cells best as the fully chelated metal complex which is less polar. Once inside the cell, the half chelated complex is required for toxicity. This half chelate may arise from the full chelate either by dissociation, or by the action of a hypothetical enzyme. Their theory is illustrated by Equation 14. Excess metal in the ambient fluid would drive the last equation to the left, reducing the amount of fully chelated oxine needed for penetration. Excess oxine within the cell would drive the last equation to the right, reducing the concentration of the toxic species, the half chelate. Other chelators in excess will also caused reversal of oxine, ostensibly by competing for metals. Block (1955) demonstrated that in addition to the ability to cnelate, enalogues of oxine must also be lipoid-soluble to be fungitoxic.

Oxine$^-$ + Metal^{+++} (Oxine) Metal$^+$
Half chelate
(Toxic within cell)

(Oxine) Metal$^+$ + Oxine (Oxine)$_2$ Metal
Full chelate
(Needed for permeation)

There are at least two discrepancies in the theory that should be noted. The first is that unchelated oxine enters cells quite readily. In fact, the theory for the reversal of toxicity by excess oxine depends on its entry into the cell. The second discrepancy is that washing with water will revive spores poisoned with oxine in a metal-deficient medium, but will not revive copper oxinate-treated spores. Washing with 0.1 *N* HCL is needed to revive the latter. These anomalies imply that unchelated oxine, which is toxic, acts in an area of the cell which does not require high lipoid-solubility to be entered. This action of oxine may well be metal-robbing. The chelated oxine acts at a site which requires high lipoid-solubility. This deeper site may well be within a microsome or the nucleus. The action of the half chelate may be the blocking of functional sites as suggested earlier.

Other workers have different views about the toxicity of oxine. Sexton (1953) considers it likely that oxine is fungitoxic because of its phenolic properties. Gale and Folkes (1957) find that oxine is one of the best inhibitors of the incorporation of glycine in disrupted staphylococcal cells. They consider that the

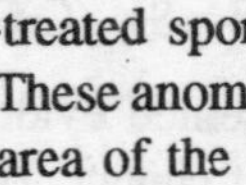

portion of oxine structurally resembles, and competes with natural components of the system they were studying. They base this conclusion on two observations: (*a*) their system is stimulated by 6-amino-4-hydroxy derivatives of benzimidazole or benzotriazole; and (*b*) their system is not inhibited by the chelator ethylenediamine tetraacetic acid.

The next heterocyclic nitrogen compound to be discussed in captan (Fig. 8.11). The discussion of captan toxicity has centered on the importance of the —$SC(Cl)_3$ group. Included here, then, will be some analogues of captan that are not heterocyclic nitrogen compounds, but do have the —$SC(Cl)_3$ group.

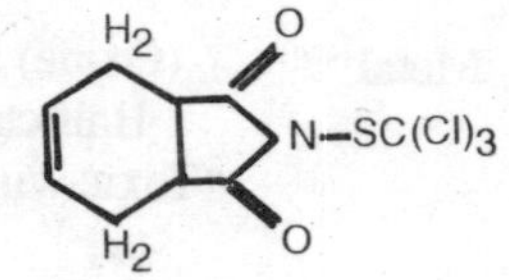

Fig. 8.11. Captan.

The fungitoxicity of the captan was first reported by Kittleson (1952). He considered the fungitoxic group to be

N—S—($C(Cl)_3$).

Horsfal and Rich (1953) presented evidence that —CO—NR—CO— could be the toxic grouping of captan. They based their conclusion on the similarity of the dosage-response slopes of captan and the highly fungitoxic antibiotic cycloheximide. Cycloheximide has the —CO—NH—CO— group but not —$SC(C)_3$. Horsfall and Rich assigned to the trichlororomethylthio group the function of getting the tetrachydrophthalimide portion of captan into the cell. *N*-nitrosophthalimidine is another highly fungitoxic compound with the —CO—NR—CO— group but no —$SC(Cl)_3$. The fungitoxicity of this compound was also used by Horsfall and Rich to support their theory. Horsfall and Rich (1954) pointed out that electronegative groups adjacent to a sulfur atom enhance activity. The electronegative carbonyl groups in captan would draw electrons from the nitrogen atom, making the *N*—S bonds less stable.

Uhlenbroek and Koopmans (1957) found that —$SC(Cl)_3$ alone did not make a compound fungitoxic. They reported that $H_3CSSC(Cl)_3$ is non-fungitoxic, $C_6H_5SCC(Cl)_3$ is slightly fungitoxic, while 2–$NO_2C_6H_4$ $SSC(Cl)_3$ is very fungitoxic. In addition, they found that —$C(Cl)_3$ could be replaced without destroying fungitoxicity completely. For example, the ED50 of $H_3CC_6H_4SO_2SC(Cl)_3$ to *Fusarium culmorum* is in the range of 0.315 to 1.0 mg./liter, very fungitoxic. Replacing the —$C(Cl)_3$ with a —CH_3 results in a compound with an ED50 in the range 1.0—3.1 mg./liter, still quite fungitoxic. Uhlenbroek and Koopmans found that the more electronegative groups on R in $RSSC(Cl)_3$, the more fungitoxic the analogue. In $RSO_2SC(Cl)_3$, however the analogues are quite fungitoxic, regardless of the electronegativity of R. They concluded that the hypothesis of increased fungitoxicity by electronegative groups adjacent to an S atom may apply for $RSSC(Cl)_3$ but not for $RSO_2SC(Cl)_3$. According to Ingold (1953), however, —SO_2R is a very strong electronegative group in itself. Changes in R, therefore, may not give good evidence for or electronegativity theory, when —SO_2R is adjacent to a sulfur atom.

Another theory of captan action was presented by Lukens and Sisler (1958). They found that captan interacts with sulfhydrul compounds, such as cysteine, to give thiophosgene

$$[Cl—\overset{\overset{S}{\|}}{C}—Cl]$$

as in Equation 15.

$$\text{Captan (N–S–C(Cl)}_3\text{)} + 2RSH \rightarrow \text{tetrahydrophthalimide (NH)} + Cl—\overset{\overset{S}{\|}}{C}—Cl + R—S—S—R + HCl$$

A clue to this breakdown is in the odor of formulated captan which is that of thiophosgene.

The thiophosegene formed may react with water (Equation 16).

$$\underset{\text{Thiophosgene}}{Cl—\overset{\overset{S}{\|}}{C}—Cl} + H_2O \rightarrow H_2S + CO_2 + 2HCl \qquad \text{(Eq. 16)}$$

Thiophosegene also polymerizes to give dithiophosgene which in turn may react with water (Eqation 17).

$$\underset{\text{Dithiophosgene}}{Cl—\overset{\overset{S}{\|}}{C}—S—C(Sl)_2} + 2H_2O \rightarrow CS_2 + CO_2 + 4HCl \qquad \text{(Eq. 17)}$$

Lukens and Sisler proposed that the toxicity of captan is produced by the highly reactive thiophosgene resulting from the breakdown of captan by sulfhydryl groups. They were also able to demonstrate this breakdown by reacting captan with yeast cells. They concluded that essentialy all of the toxicity of captan to yeast can be accounted by the —S—C(Cl)$_3$ group. Thiophosgene produced *in situ* would ostensibly poison by combining with free sulfhydryl, amino hydroxyl, and perhaps carboxyl groups. In addition, thiophosgene could form

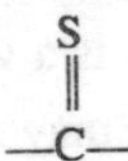

bridges between properly spaced groups within the cell.

The production of thiophosgene is undoubtedly of importance to captan activity. Whether or not the captan acts simply as a thiophosege producer remains doubtful. If, as Uhlenbroek and Koopmans (1957) reported, some —S—C(Cl)$_3$ compounds are nontoxic, and if replacing —C(Cl)$_3$ does not destroy fungitoxicity, than thiophosgene production may not be the whole stroy of —S—C(Cl)$_3$ compounds. Captan, which loses its chlorine atoms readily, may well form the *N*-chlorotetrahydroph, thalimide as a transitory interemediate, analogous to diisothiocyanates formed in nabam breakdown. That the *N*-chloro derivative may form is supported by Orwoll (1954). He presented the following possibility (Equation 18) for the breakdown of a captan-type molecule.

$$Ar—SCCl_2 \rightarrow CSCl_2 + ArCl \qquad \text{(Eq. 18)}$$

Although *N*-chlorotetrahydrophthalimide is too unstable to test, its analogue, *N*-chlorosuccinimide, is a well known and highly active germicide. The breaking of the N—S bond would leave the N position free to combine with any reactive grouping within the cell. Captan, then may poison by releasing highly reactive tetrahydrophthalimide radicals, thiophosgene, and chlorine.

So much for the heterocyclic nitrogen compounds, a particularly rich source of fungitoxicants.

D. Quinones

Quinones are among the most powerful of the agricultural fungicides. The two most commonly used in practice are both chlorinated compounds. These are 1, 4-tetrachlorbenzoquinone (chloranil), and 2, 3-dichloro-1, 4-napthoquinone (dichlone).

Chloranil has found its use primarily as a seed protectant. As a foliage fungicide it usually fails to protect, being readily photolysed to nontoxic products.

Dichone, much less readily photolyses, is widely used as a foliage spray. It has been particularly successful against apple scab. The potency of the quinones may be illustrated by the usage of dichlone for controlling apple scab. Most protective fungicides for apple scab control require 1 to 3 lb. of formulation per 100 gallons. Dichlone is used successfully at a dosage of 1/8 to 1/4 lb. per 100 gallons.

Owens (1953a) points out a number of ways by which quinones may poison spores. As *alpha, beta*-unsaturated compounds they may combine with sulfhydrul groups (Equation 19).

They may also interact with amino groups.

Quinones may also poison by rerouting electron transfer systems as described in an earlier section.

McNew and Burchfied (1951) found the following order of increasing fungitoxicity: anthraquinone < benzoquinone < phenanthraquinone < naphthoquinone. Halogenation, while decreasing water solubility and phytotoxicity, increases fungitoxicity. Chlorination improves fungitoxicity the most, bromination next, iodination least. Alkylation ortho to carbonyl groups reduces fungitoxicity, presumably by steric hindrance.

$$\text{(quinone)} + HS-R \xrightarrow{1/2\ OX} \text{(quinone)}-SR + H_2O$$

A point of interest is that the quinonoid structure may not be necessary to the activity of these compounds. For example, Owens (1953) reports that 2, 5-dichlorohydroquinone is almost as toxic as chloranil. Of course, data of Rich and Horsfall (1954) suggest that the hydroquinone may be oxidized to the quinone by fungi.

E. Phenolic Compounds

Phenolic compounds have been of interest as germicides since the introduction of carbolic acid into surgery by Lister in the last century. Phenols as fungitoxicity or germicides have been particularly useful in medicine and industry rather than in agriculture. One of the most widely used germicides in medicinal preparations is hexylresorcinol. Various derivatives of cresol and phenylphenol are used as wool preservatives. Sodium dinitro-*o*-cresylate has been used as a spray on orchard floors to kill overwintering forms of fungi which cause fruit diseases.

One of the significant contributions to the structure-fungitoxicity relationships of phenols was made by Shirk (1954). He found that alkylating both positions ortho to the phenolic —OH greatly reduced fungitoxicity. In addition, *o*-chloro-6-alkylphenols are less fungitoxic than the corresponding *o*-alkylphenols. Shirk explains this effect as steric hindrance. He bases his suggestion on the report of Stillson et. al. (1945) who found that phenols substituted 2, 6 with large alkyl groups gave reactions not at all like typical phenols. These they called hindered phenols. Interfering with the reactivity of the phenolic —OH, then, profoundly affects fungitoxicity.

Another group of phenolic fungicides of interest are the bisphenols. The germicidal action of these compounds was first reported by Bechold and Ehrlich (1906). Marsh et. al. (1945), and Marsh and Butler (1946) found bisphenols to be highly effective mildew preventies on cloth. Their best material was 2, 2′-methylenebis (4-chlorophenol) also known as G-4, preventol GD and dichlorophene (Fig. 8.12). Goldsworthy and

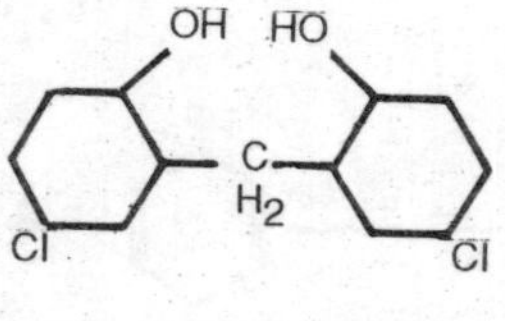

Fig. 8.12. 2, 2′-Methylenebis (4-chlorophenol).

Gertler (1949) reported this compound to be an excellent foliage fungicide for peach trees.

The corresponding thibisphenols are also effective funicides. Here, the methane bridge is replaced by a sulfur atom. Pfleger et. al. (1949), found that the thiobisphenols are highly toxic *in vitro* to fungi causing human diseases. Schraufstätter et. al. (1949) reported 2, 2′-thiobis (4-chlorophenol) as a particularly good antimycotic. Horsfall and Rich (1950) showed Schraufstätter's compound to be an excellent foliage protectant against apple scab. Unfortunately, it damaged the fruit.

Cade and Gump (1954) summarized the results of structure-activity research with bisphenols. Activity is increased by halogenation of each ring in the 4-positions. Any more halogenation decreases activity. The phenolic —OH's are most effective ortho to the bridge linking the rings. The type of bridge between the two rings is not too important. The bridge must be small enough, however, to prevent spreading the hydroxyls too far apart. This latter evidence indicates that chelation may play a part in the toxicity of the bisphenols.

There are many other kinds of fungicides, such as salicylates and benzoates, that have been considered phenolic toxicants. The large group of phenolic toxicants are usually considered as nonspecific, protoplasmic poisons.

F. Halogenated and Nitrated Aromatic Compounds

The halogenated aromatic compounds are usually toxic because they readily release nascent halogens. This is undoubtedly the mechanism of action of such compounds as chloramine-T and *N*-chlorosuccinimide. The freed chlorine may act to form oxidizing $HClO^-$ as in the case of the hypochlorites. Another kind of action may be a direct halogenation of a metabolite to form an antimetabolite.

Nitrated aromatic compounds most potent as fungicides are usually either halogenated or they are phenols. One of these compounds, 2, 4-dinitro-*o*-cresol was mentioned in the previous section. Another is 2, 4-dinitrophenol, a powerful uncoupler of phosphorylation. Of the halogenated nitro-aromatics, one of the most powerful fungicides is 1-fluoro-3-bromo-4, 6-dinitrobenzene. The chlorinated nitrobenzenes have been of continuing interest since they were first described as fungitoxicants. Pentachloronitrobenzene and tetrachloronitrobenzene have found use both as soil fungicides and for the control of *Botrytis* sp. on plant parts. These compounds prevent sporulation, and are vapor-phase fungistats which do not prevent spores from germinating but do stop hyphal growth. Horsfall (1956) suggests that compounds of this type are antimitotics. It may be that pentachloronitrobenzene and tetrachloronitrobenzene act specifically as competitive inhibitors for inositol, a required growth substance for many fungi.

G. Other Organic Fungitoxicants

The list of fungitoxic organic compounds is a long one. They cannot all be mentioned in one short chapter. Such fungitoxicans as quaternary ammonium compounds were mentioned in the section as surfactants. These have been tried for the control of plants diseases.

Not yet mentioned are the aliphatic fungitoxicants. These are of not too much importance and most probably act as nonspecific, physical toxicants. The organometallic fungicides have been of particular interest, so should not be omitted. Suffice it to say that many organometallic compounds, such as the organic mercurials, are more powerful than the inorganic compounds of the same metals. Of interest in this regard is tin, not very fungitoxic in inorganic combinations, but producing some highly fungitoxic organic compounds.

IV. FASHIONING FUNGICIDES

Success of Screening Programs

The artistry of the synthesizing chemists combined with the carefully designed screening programs of the biologists have given us a wealth of new and highly effective fungicides in recent years. The methods used for screening are as various as the laboratories so engaged. Although the selection of compounds to be screened is usually empirical, the tests to which the compounds are subjected are based on knowledge of plant diseases. Regardless of the method, there is very little question that most of the commericial fungicides now in use would have been selected by any of the presently operating screening programs. The screens work.

Concomitant with the selection of candidate fungicides, the screening programs have resulted in masses of structure-activity data. These data have been carefully scrutinized for "philosopher's stones" which will predict fungitoxic configurations. The literature abounds with such studies. These have been of use in predicting the activity of closely related compounds, and in a few cases have predicted the activity of untested configurations.

9

EFFECT OF NEMATOCIDES

The science of plant pathology was concerned primarily with the fungi during the first 50 years. Then the bacteria had a short period of prominence. This did not last long however, because fungi were still more numerous as pathogens. Bacteria were pushed into the background, too, by the viruses which have been demanding increasingly more attention.

The nematodes are not "newcomers" as plant pathogens, the wheat nematode was known as a plant pathogens in 1744 and other important species were described in the 19th century. Present-day general acceptance of nematodes as important plant pathogens was arrived at slowly. Numerous reasons could be advanced to account for the seeming reluctance of plant scientists to embrace nematodes as plant pathogens. Perhaps the fact that they are animals and have been the concern of the zoologist, particularly the parasitologist, has tended to place the plant pathogenic nematodes in the unique position of being unclaimed, until recently, by any organized scientific discipline.

However, throughout this long period of general apathy towards the plant parasitic nematodes, their cause was proclaimed far and wide by a handful of devoted specialists in nematology who were concerned primarily with the importance of nematodes as destroyers of cultivated plants.

As the importance of many types of nematodes as plant pathogens was recognized, plant pathologists became more concerned with them. The result is that we have a new race of plant pathologists—those concerned with animals that are plant pathogenic. Moreover, we still have the zoologist specializing in nematology and contributing much of the fundamental information concerning systematics, morphology, physiology, and ecology of the soil inhabiting nematodes. The problem is of great magnitude and presently requires the additional knowledge that will result from the cooperation for these animals and plant scientists.

1. THE PROBLEM OF PROVING PATHOGENTICITY

Nematologists have struggled again over the same old rough ground traversed by the mycologists, bacteriologists, and virologists. How can one prove that a nematode is a pathogen? Nematologists inherited agar techniques and petri plates from their predecessors but little else that was significant in the search for the elusive proof.

The extreme difficulty of obtaining pure cultures of nematodes has made to proof most difficult. Without proof, many plant pathologists have been loath to accept the pathogenicity of many species. They have accepted constant association as sufficient evidence for pathogenicity of such obvious nematodes as the root knot nematode and the golden nematode.

Nematocides have contributed enormously to the proof needed by the nematologists, who were not able, at least until recently, to culture the nematodes, add them to plant, produce the disease, and reisolate the original nematode. The use of nematocides to kill them has given a type of inverted proof. Constant association made this reasonably easy. Nematocides killed off the nematode and, thus, interrupted the constant association. This was nearly as good proof as Koch demanded. At least it was good enough to set many men baying down the trail of nematodes as plant pathogens. The subject grows apace.

II. WHAT A NEMATOCIDE IS

The term nematocide is restricted in this discussion to chemicals with nematode-killing properties that have been used or are currently used to reduce populations of plant pathogenic nematodes.

In addition to nematocidal activity, useful nematocides must possess special properties with regard to phytotoxicity, diffusion in soil, water solubility, vapor pressure, and cost. The chemicals that are here considered as nematocides are not specific in their toxic action. They usually kill all of the nematodes that are present in the area of their activity. Some nematocides are fungicidal, insecticidal, and herbicidal as well and could equally well be referred to as and biocides since they partially or completely sterilize the soil.

In disease complexes involving nematodes and fungi or bacteria there may be secondary benefits from the application of nematocides. Generally, too, of benefits of treatments with nematocides are measurable in terms of increased crop yield, quality, and better plant growth. In practice, it is expected that the use of nematocides will increase the profit to the user. Some investigators suggest that a soil fumigation treatment with a nematocide should increase yield or quality of the product sufficiently to return to the grower an amount equal to twice the cost of the treatment. Taylor (1951) says that it is no economical to fumigate soil for nematode control unless the gross value of the crop is at least $375 per acre. These are matters that can only be decided by individual growers, since there are many economic factors that determine whether or not a nematocide is profitable. There are situations where a nematocide is not profitable even though it more than doubles the yield of the crop, as in California when sugar beet yields have been doubled but the yield on treated land is still not an economic level. In other situations there may be no significant increase in total yield but the quality is increased to profitable level. This has been evident in such crops as carrots and potatoes. One of the attractive features of nematode control by nematocides is that the benefits of the treatment are immediately obvious to growers in terms of net profit. This is not doubt responsible for the fairly rapid adoption of nematocides once they became available at relatively low cost.

III. HISTORICAL

The application of chemicals to the soil to control plant parasitic nematodes was undertaken by a number of investigators because crop failures were associated with the presence for these pests. Kühn (1881) attempted control of the sugar beet nematode, *Heterodera schachtii* Schmidt, 1871, with a number of chemicals and concluded that the most promising was carbon bisulfide. However, this chemical was not entirely satisfactory because of the high cost of the treatment and the relatively unsatisfactory results. It is a likely that Kuhn used carbon bisulfied because of its then widespread use against grape *Phylloxera*, Neal (1889) applied carbon bisulfied to the soil as well as numerous combinations of tobacco dust, potassium sulfate, potassium sulfide, sodium hyposulfite, sulfur, and caustic lime in experiments of control root knot nematode in Florida. He also applied carbon bisulfide to living plants which were killed by the chemical because it is highly phytotoxic. Neal did not obtain satisfactory results with carbon bisulfide probably because he allowed an interval of only one day betwen treatment of the soil and planting. He did, however, indicate that tobacco dust was the most satisfactory of the treatments he tried.

It is apparent from the literature that until 1940 numerous chemicals were investigated as possible nematocides for the control of nematodes. Chloropicrin appears to have been the only satisfactory chemical tested prior to 1940 for the treatment of nematode infested soils on a commercial scale.

Chloropicrin was first reported to be an effective nematocidal chemical for use in soil by Matthews (1919). Subsequently, Johnson and Godfrey (1932) reported excellent control of root knot nematode in pineapple soils. This chemical is not widely recognized for its nematocidal, fungicidal, and herbicidal properties. As a nematocide, it was once widely used in pineapple soils in Hawaii where satisfactory increases in yield were obtained. Chloropicrin has a fairly high vapor pressure (Table 9.1) and it is most effective when confined to the soil by impervious covers, by water seal, or by the use of high dosages. The high cost of soil treatments with chloropicrin (approximately $150 to $200 per acre) has tended to restrict its use to crops of high value per acre. Taylor and McBeth (1941) developed the spot treatment method of applying chloropicrin and this made it economically feasible to treat large acreages of melons where treatment of the planting site reduced the cost to approximately $5 per acre.

Methyl bromide is sometimes used as a soil fumigant to control nematodes. Richardson and Johnson (1935) reported it to be an effective nematocide and subsequent workers have reorted the conditions for its most effective use. The chemical has had rather restricted use as a nematocide except for special crops of high value but is very effective in situations where fungicidal and herbicidal qualties are desirable. Methyl bromide has a high vapor pressure and a low boiling point, it can be effectively used only if the vapors are confined to the soil by gas impervious covers. The chemical is highly toxic to man and extreme care must be exercised to avoid breathing the vapors. A small amount of chloropicrin is frequently added to methyl bromide so that it will be more easily detected by persons using it as a fumigant.

TABLE 9.1

Physical Properties of Nematocides

Nematocide	*Boiling point (°C)*	*Vapor pressure (25 °C)*	*Specific gravity*	*Solubility (gm/100 ml. water)*	*Physical state*
Chloropierin	112	24.0	1.65	.20	Liquid
Methyl bromide	45	1824	1.73	0.09	Gas
D-D Mixture	<95 – 150>	40.4	1.2	0.27	Liquid
Ethylene dibromide	131.6	12.0	2.17	0.43	Liquid
1, 2-dibromo-3-chloropropane	193	0.9	2.08	0.15	Liquid
Sodium *N*-methyl dithocarbamatic	—	—	—	72.2	Solid
VC-13	124–130 (0.25 mm.)	—	—	0.245 (mg./1.)	Liquid

The era of modern nematocides had its beginning with the discovery by Carter (1943) of the nematocidal properties of 1, 3-dichloropropene-1, 2-dichloropropane mixture (D-D mixture). This was the first relatively cheap nematocide that could be used to control nematodes in soil at a picrin and methyl bromide, D-D mixture, because of its lower vapor pressure, does not require the use of impervious covers or water seals. In 1945, Christie reported that ethylene dibromide had given excellent control of root knot nematode in the soil in preliminary tests. Subsequently, this chemical and D-D mixture have been widely used to control nematodes in the soil.

Unlike many of the previous chemicals that were used to control nematodes, D-D mixture and ethylene dibromide have very little effect on soil fungi at the dosage rates ordinarily employed to control nematodes in the field. Both chemicals do, however, kill soil animals and both have been used as soil insecticides. Ethylene dibromide was widely used to control wireworms until the advent of the chlorinated hydrocarbon insecticides.

Since the discovery of D-D mixture and ethylene dibromide, one additional chemical has been developed that is now widely used, this is 1, 2 dibromo-3-chloropropane reported by McBeth and Bergeson (1955). Later in 1955 Allen et. al. reported the chemical to be an effective nematocide based on experiments conducted in 1953 and 1954. This material presents a new aspect, low phytotoxicity. Thus, it can be safely applied to living plans.

Historically, the discovery of D-D mixture and ethylene dibromide and their subsequent wide use as low cost effective nematocides has had a very profound influence on our knowledge of plant pathogenic nematodes. Many nematodes ectoparasite on roots have been discovered by the use of one or the other of these nematocides in fields where crop growth and yields were reduced from unknown causes. In addition, the use of these materials has made possible a more accurate appraisal of the damage that should be attributable to nematodes. The spectacular increases in yield that have frequently resulted when nematodes are controlled have served to make growers aware of the necessity of employing nematode control procedures. The net result of the use of these materials has been the awakening of a widespread interest in nematodes as plant pathogens, recognition that nematodes are frequently important in certain disease complexes, and renewed activity in the study of nematodes.

IV. THE EFFECT OF NEMATODE CONTROL ON THE CROP

Nematocides are usually applied in situations where control of nematodes will result in increased quantity or quality of yield of the harvested portion of the crop. It is not possible to cite all instances of increased yields that have been reported in the literature. The fact that the use of nematocides has increased many fold during the past 15 years is ample testimony to the effectiveness of nematocides in increasing crop yields. Thorne and Jensen (1946); Raski et. al. (1953); Allen et. al. (1955); Lear and Thomason (1956); Stark et. al. (1944) and numerous other investigators have reported the increased yields resulting from the use of nematocides.

There is no method, aside from experience, that permits an accurate forecast of the yield increases that can be obtained from nematode control treatments. Over a period of 10 years when all plots were carefully selected on the basis of previous nematode damage from root knot nematode and for soil types favorable to nematocidal treatments, the yield of lint cotton in California tests has varied from no significant increase to more than one bale per acre with both broadcast and row placement treatments. The percentage increase does not appear to be correlated with the yield obtained on untreated soils. For example, Allen and Raski (1950) obtained yield increases of 0.6 to 1.0 bale of lint cotton when the yields on untreated plots were only slightly more than one bale per acre. In 1957 preplanting row placement treatments with 9 gallons of D-D mixture per acre resulted in a yield of 3.50 bales of lint cotton per acre on land that produced 2.57 bales of lint cotton without treatment. Although satisfactory yields are sometimes obtained on land which is heavily infested with nematodes, this should not necessarily be the final criterion in determining the necessity of controlling nematodes. This is so, because of the variety of soils, and cultural practices that are encountered in the field. If it is known that nematodes are present, then exploratory treatments should be undertaken to determine the advisability of applying a nematocide.

Nematocides applied for the express purpose of securing the maximum reduction in populations and maximum yield that are possible are applied as area treatments. In these treatments, an attempt is made to kill all of the nematodes in the upper layers of the soil. Conventional treatments of this type involve the injection of the volatile nematocides by shanks spaced 12 inches apart and to a depth of 8 inches in the soil. D-D mixture at 20 gallons per acre, 83% ethylene dibromide at 4–6 gallons per acre, or 1, 2-dibromo-3-chloropropane at 1 to 2 gallons per acre are the dosage rates ordinarily used to control root knot nematode in sandy or sandy loam soils. Because of the extreme range of soile types it is not possible to predict the exact results that can be obtained from a specific treatment. However, the application of 20 gallons of D-D mixture per acre to a cotton field in California in 1946 resulted in nearly 100% control of root knot nematode larvae to a depth of 5 feet. Examination of the roots of cotton grow subsequently in this field also indicated a high degree of control. This treatment satisfactorily controlled the nematodes for two crops of cotton. However, this is an exceptional circumstance as indicated by Allen et. al. (1955). Row treatments with 9 gallons of D-D mixture per acre increased cotton yields by 0.3 bale of lint cotton per acre on plots treated the previous year with 20 gallons of D-D mixture per acre. More striking results have been obtained in tests on sugar beet nematode by Thorne (1952 where there was complete crop failure of beets planted on soil that had been treated and had grown a crop of beets the previous year.

There are nematocidal treatments that are directed toward eliminating all of the plant pathogenic nematodes in the soil. Such is the case in the preplanting treatments of soils for perennial crops such as citrus, walnut, apple, cherry, peach, grape. Here it is necessary to apply high dosages so that kills are obtaiend to considerable depths in the soil. However, even in sandy soils kills are only obtained to a depth of about 8 feet.

Row placement treatments are a variation of the spot treatments suggested by Taylor and MacBeth (1941). They are applied to protect the root system of the plant from early attack by large number of nematodes. The degree of control is not as high as is obtained with area treatments. and usually the increases in yield are not as great. However, a lesser amount of the fumigant is applied so that the net profit obtained from row treatments are usually about equal to those from area treatments. The amount of nematocide required for row placement treatments varies with the row spacing. In cotton where row spacings are 38 to 40 inches, satisfactory control of root knot nematode can be obtained by the application of 7 to 9 gallons per acre of the dichloropropene nematocides, 2.5 to 3 gallons per acre of 83% ethylene dibromide or 0.5 to 0.75 gallons per acre of 1, 2, dibromo-3-chloropropane. The amount required is less if row spacings are greater than those indicated. Row placement treatments many be applied with one, two, or three chisels spaced at 12 inches depending on the crop to be grown.

A. Nutritional Effects

Plants injured by nematode infestations exhibit symptoms that are characteristic of nutritional deficiencies. The poor growth and symptoms of sugar beets infested with sugar beet nematode were once thought to be caused by a potassium deficiency. However, such plants failed to respond to soil applications of potassium, or other fertilizers and it was later found that the primary cause of the trouble was sugar beet nematode. Nematode infested plants are frequently deficient in certain essential elements and it is not surprising that control for nematodes sometimes results in the correction of deficiency symptoms. One of the corrective measures that is first used by growers on crops that exhibit poor growth and yellowing of the foliage is the application of additional amounts of fertilizer. This procedure has been undertaken with little success in many problems that involved nematodes. When the problem has been corrected by the control of nematodes, the plant grown on treated soil appear to be greatly stimulated and symptoms of nutritional deficiency have been alleviated. This situation has caused some investigator to postulate that nematocides applied to the soil cause the release of some essential nutritional elements. Although there does exist the possibility of some ion exchange as the result of the application of nematocides, such claims have not been substantiated. The so- called "increased growth response" of plants grown on soil treated to control nematodes is now recognized to be due largely to the control of nematodes and the subsequent development of normal, functional roots by the plant.

The effect of soil fumigation with D-D mixture and chloropicrin on nitrification and growth of pineapple has been studied by Tam (1945). He found that these nematocides suppress nitrification, that ammonia accumulates in the soil, and that pineapple plants are able to obtain nitrogen from the ammonia. He indicated also that chloropicrin suppresses nitrification for a longer period of time than D-D mixture because of its greater toxicity to nitrifying bacteria. Increased amounts of nitrogen are also present in the leaves of plants grown on treated soil.

Tarjan (1950) reported that boxwood infested with *Pratylenchus* spp. is deficient in several elements and there was some indication of an increase in the amount of sodium in the leaves. Oteifa (1952) found that lima beans infested with root knot nematode have lower amounts of nitrogen, phosphorus, calcium, magnesium, and potassium. He also indicated that there is differential uptake of these elements in the presence of nematodes. Sher (1958) found that the leaves of roses infested with the root lesion nematode, *Pratylenchus vulnus* (Allen and Jensen, 1951), contain lower amounts of iron, copper, and magnesium. The foliage of the infested roses exhibited symptoms of deficiency discease. Rose plants not inoculated with the root lesion nematode and growing in the same soil as inoculated rosese do not show symptoms of deficiency. It might be anticipated that one of the side effects of controlling parasitic nematodes would be restoration of normal nutrition of plants if they are growing in soil with sufficient available amounts of required elements. This seems to be so in the field since plants grown on treated soil are usually normal in appearance and make vigorous growth.

B. Water Relationships

That nematode infested plants suffer from water deficiency was observed many years ago in sugar beets infested with sugar beet nematode. Infested plants frequently exhibit severe wilting during the daytime and recover at night. In cases of heavy infestations, this condition of wilting becomes permanent and the plants do not recover regardless of the application of large amounts of irrigation water. This is an extreme example and does not occur to the same extent in other crops. In conducting trials in California with nematocides to control root knot nematode infestation for cotton we have observed that cotton plants growing on untreated plots show evidence of the need for irrigation in advance of plants on treated plots. In these experiments., if irrigations were applied on the basis of the need of the plants on the untreated plots, then plants on treated plots were irrigated too frequently. This results in a lowered yield on the treated plots due to the increased vegetative growth induced by overirrigation. It has been the experience of growers that the use of nematocides on root not nematode infested soils frequently allows a reduction in the number of irrigations required when the same soils were not treated for nematode control. It is sometimes stated by growers that nematodes not only reduce yields, but they also increase the production costs because of the excessive amounts of fertilizers and water that are required to grow susceptible crops on infested land.

C. Disease Complexes

The associations of nematodes with certain fungal and bacterial diseases of plants have been suggested by a number of early authors. Gilbert (1914) suggested that *Fusarium* wilt of cotton causes greater crop losses when root knot nematode is present. Hastings and Bosher (1938) found that meadow nematode and *Cylindrocarpon radicicola* Wr. cause greater damage in combination than is caused by either organism alone. Young (1939) presented evidence indicating that root knot nematode decreases resistance of many tomato varieties to *Fusarium*. To the present time there have been numerous contributions to the subject of disease complexes involving nematodes and other organisms. Some of the best known of these associations are as follows: meadow nematode and brown root rot of tobacco; meadow nematode and root rot disease of small grains; *Aphelenchoides ritzema-bosi* and *Corynebacterium fascians* in cauliflower disease of strawberry; sting nematode, *Belonolaimus gracilis*, and *Fusarium* wilt of cotton; reniform nematode, *Rotylenchulus reniformis*, and *Fusarium* wilt of cotton; root knot nematodes, *Meloidogyne* spp., and black shank of tobacco; root knot nematode and Granville wilt of cotton, nematodes and bacterial wilt of carnations tobacco stunt nematode and *Fusarium* wilt of tobacco; root knot nematodes, *Pratylenchus* spp., *Tylenchorhynchus claytoni* and black shank of tobacco, root knot nematode and *Rhizoctonia disease* disease of cotton.

One of the most significant development in these disease complexes involving nematodes is the discovery that certain disease resistant varieties lose their resistance when nematodes are present along with the fungus or bacteria. Equally important is the fact that if nematodes are controlled by nematocides the plants do not lose their resistance. It is also of interest that some of these relationships between nematodes, pathogenic fungi, and bacteria, have been discovered through the use of nematocides. Certain of these diseases have been controlled by the use of nematocides before there was evidence of the cause of the disease. These instances have sometimes resulted in the discovery that nematodes are the primary cause of the disease and that other associated organisms are secondary, e.g., brown root rot of tobacco. Perhaps the most far-reaching of the developments in this field is the discovery that nematodes are sometimes responsible for breaking of resistance and the resistance can be preserved by the use of soil treatments that reduce or control nematode populations.

V. FOLIAR AND SEED-BORNE NEMATODES

Most pathogenic nematodes live in the soil and invade the roots of plants. On that account, most nematocides are soil treatments.

Still, there are a few foliar and seed-borne nematodes and they will be treated briefly before going into the main topic of soil-borne nematodes, and soil nematocides.

A Foliar Nematodes

Several species in the genus *Aphelenchoides* are pathogens on the foliage of plants. They are ordinarily not present in large numbers in the soil, even in soil around infested plants. Control of these species with nematocides involves treatment of the foliage with chemicals that have relatively low phytotoxicity. A unique characteristic of several of these species is their ability to be ecto-or endoparasitic depending on the host plant. *A. ritzema-bosi* is endoparasitic in chrysanthemum, but if transferred to strawberry it becomes an ectoparasite. It has been observed to be both endoparasitic and ectoparasitic on gooseberry. *A. fragariae* is ectoparasitic on strawberry but becomes an endoparasite in Croft lily and ferns. As ectoparasites these nematodes ordinarily are located in tightly folded leaf and flower buds. When the leaves unfold and are exposed to drying, the nematodes soon disappear from the exposed surfaces.

Practical control of foliar nematodes with chemicals has been obtained with relatively few materials. Chemicals are not ordinarily used to control any of the *Anguina* spp. that attack cultivated plants.

The use of sodium selenate as a systemic nematocide to control foliar nematode infestations in chrysanthemums has been reported by Dimock (1944). Tarjan (1950) investigated the use of this material as a soil treatment to control *Pratylenchus* spp. infesting boxwoods and indicated that root populations are significantly reduced.

Practical control of foliar type nematodes by the use of chemical sprays became possible with the development of the organic phosphates as insecticides. Raski and Allen (1948) used parathion as a foliar spray on strawberries infested with *Aphelenchoides fragariae*. Populations were reduced to a low level, but satisfactory control was not obtained in tightly folded leaf buds. Dimock and Ford (1950) obtained "spectacular" control of the foliar nematode of chrysanthemum, *Aphelenchoides ritzema-bosi*, with parathion sprays. Systox sprays have been found to be effective in experiments on the stem and bulb nematode in daffodils by Feder (1952) and in alfalfa by Bergeson (1955).

B. Seed-Borne Nematodes

Nematodes carried as seed contaminants, particularly stem and bulb nematode on onion, teazle, and clover seed can be successfully controlled by fumigation with methyl bromide, T. Goodey (1945) and J.B. Goodey (1949). However, methyl bromide fumigation is not effective against nematodes imbedded in plant tissues or within seeds or galls. This is undoubtedly due to poor penetration and to the fact that nematodes within galls (wheat nema) are usually very inactive at the time of treatment. 3-*p*-chlorophenyl-5-methyl rhodanine has been reported to control *A. besseyi* infestations of rice seed. Hot water treatments have been much more satisfactory for the treatment of bulbs and corms infested with nematodes.

VI. SOIL-BORNE NEMATODES

Since most plant pathogenic nematodes live in the soil and attack the plants from the position, our best information deals with soilborne nematodes. As a basis for understanding how nematocides act and how nematocides are used, we must first discuss the biology of the nematodes themselves.

Plant nematodes, with few exceptions, are unable to complete their life cycle except in the presence of higher plants. Their life cycles are relatively simple and consist of egg, larval, and adult stages. All species, that have been thoroughly studied, have been found to undergo one larval molt in the egg, so that the youngest active stage found in the soil or within plant tissue is usually the second stage larva. There are ordinarily three larval stages outside the egg. These larval stages and the adult are found in the soil in some species of plant parasites while in other species only the second stage larvae are found in the soil. Because of differences in parasitic habit and certain specializations exhibited by some groups of plant pathogenic nematodes it is desirable to classify them into a number of groups and consider each group separately.

A. Ectoparasites of Roots and Underground Parts of Plants

The ectoparasite nematodes are characterized by having all stages present in the soil around the roots of host plants. These nemas feed upon plant cells by means of the stomatostyle or odotostyle depending on the particular parasite concerned. Ordinarily, the ectoparasites of the underground parts of plants do not enter the roots. They are able to move from one feeding site to another and they are active nematodes in the larval and adult stages. These nemas are also characterized by the fact that they lay their eggs singly in the soil and the eggs are not known to be protected by any specialized structures other than the egg shell.

In considering control of ectoparasites of underground plant parts we can conclude that it is unlikely that there will be present, in the soil in the absence of a host, large numbers of eggs. The majority of the soil population will be adults and larvae. So far as is known these stages shown no particular resistance to the action of nematocides. The ectoparasites are regarded as easily killed by present-day nematocides. An exception is the stubby root nematode according to Perry (1953). It has been reported by Todd and Furgeson (1956) that ethylene dibromide is superior to D-D for control of tobacco stunt nematode and by Good and Parham (1957) that ethylene dibromide is more effective against sting nematode.

Ectoparasitic nemas should theoretically be easier to control than endoparasites with postplanting applications of nematocides because they are not protected by plant tissues and because eggs deposited in the soil are without specialized protective structures. Included in this group are nemas in the following genera: *Trichodorus, Tylenchorhynchus, Rotylenchus, Helicotylenchus, Belonolaimus, Dolichodorus, Xiphinema, Longidorus, Criconemoids, Paratylenchus, Hemicycliophora,* and *Criconema.*

B. Vagrant Endoparasites

Nematodes in this group are characterized by their endoparasitic habit as well as by the fact that larval and adult stages retain their ability to move actively in soil and roots. The females and males do not, at any time, become sessile or saccate. In some respects the vagrant endoparasites are similar to the ectoparasitic group in that they are able to leave the feeding site. In addition, eggs are deposited singly. Unlike the ectoparasities the vagrant endoparasitic species deposit the majority of their eggs with the host tissues. Thus, it is possible for the individual nematode to spend its entire life span within the host and never be exposed to the soil environment outside the host tissues.

The presence of endoparasitic nematodes within host tissues has an important implication with respect to the application of nematocides to soil. It is well established that nematodes in undecomposed root, bulb, corm, or tuber tissues are protected from the toxic action of the various nematocides. In order to kill the nematodes and their eggs satisfactorily within the plant tissues, greater volumes of fumigant are usually required. It is, therefore, extremely important to recognize the type of nematode involved and to undertake preplanting treatment only after the infested roots of previous crops have decomposed or to use the higher dosage rate.

Species in the genera *Pratylenchus* and *Radopholus* are the most commonly encountered vagrant endoparasites. The nemas in these genera are frequently more difficult to control in ordinary field preplanting treatments than the infective larvae of the root knot nematodes.

There is a marked difference in ectoparasitic and vagrant endoparasitic nematodes with respect to their control with postplanting application of nematocides. The protection afforded the endoparasites by the host tissues adds materially to the difficulty of securing adequate population reductions. It has been observed that the vagrant endoparasitic forms that are outside root tissues are readily killed by postplanting treatments but those individuals inside plant tissues ordinarily survive treatment unless the root tissues are also killed by the nematocide. It is thought that nematocides with relatively low phytotoxicity and long residual action are more effective because they remain in the soil and are toxic to nemas that leave plant tissues. Experimental results to the present time indicate that postplanting treatments are generally of little value when applied to heavily infected mature plants.

C. Exposed Endoparasites (Noncyst-Forming)

The placement of several forms in this classification is questionable but they do differ in their parasitic habit from those nemas previously discussed as well as those discussed in the following sections. These nemas undergo rather extreme morphological changes in their development. The females becomes saccate and are unable to move from the site of feeding after they have penetrated into host plant tissues. The exposed sessile endoparasites never completely penetrate into the host tissue. The infective stage penetrates only partly into the host with the posterior portion of the body remaining outside the root. Males of these species do not penetrate plant tissues.

In addition to their sessile habit these nemas deposit their eggs into a gelatinous matrix which serves as a protective device against drying and possibly affords some protection from parasitic and predaceous soil organisms. It is also known that the gelatinous matrix affords some protection from the toxic action of nematocides. However, unlike the nemas in the previous group the exposed sessile endoparasites are not at any time completely protected by plant tissues, and are exposed to the toxic action of nematocides regardless of whether they are present on roots at the time of treatment. They are generally regarded as more susceptible to the action of nematocides than those nematodes that are completely protected by root tissues. The major characteristic of this group is the presence of the gelatinous matrix and the production of a relatively large number of eggs. At certain times most of the population of nemas in this group occur as eggs. Unsatisfactory results may be obtain under such conditions unless the nematocide used has high ovicidal activity.

The two best known species in this group are the citrus nematode, *Tylenchulus semipenetrans* Cobb, 1913, and the reniform nematode, *Rotylenchus reniformis* Linford and Oliveira, 1940. Some reports indicate that postplanting treatments of crops infested with these nematodes have been more successful than of crops infested with endoparasites that are completely embedded in root tissues. Experimental results with such treatments on citrus infested with citrus nematode have been promising using 1, 2-dibromo-3- chloropropane. This material and ethylene dibromide have been used successfully as side dressing treatments on pineapple infested with reniform nematode. In the latter case the side dressing treatments are made at intervals following planting of the crop on beds receiving preplanting treatments with D-D mixture or ethylene dibromide.

D. Nonexposed Sessile Parasites

In this group are included the root knot nematodes, *Nacobbus* species, and *Ditylenchus radicicola*, a pathogen on grasses. These nematodes are characterized by complete penetration of the infective stage into the host plant. In many instances the nematodes mature and produce eggs that are wholly or partly embedded in plant tissue. In other cases, e.g., root knot nematodes, the female is inside the plant tissue and the eggs are deposited in a gelatinous matrix that is wholly or partly outside the tissue. Additionally, species in these genera cause the formation of galls which completely or partially enclose the nematode. Under ordinary circumstances control of these nematodes is directed against the infective stage or stages in the soil. The second stage larva is the infective stage of root knot nematode and this probably is also the case with *Nacobbus*. Abundant experimental evidence in the literature indicates that when eggs and larvae occur in the roots, satisfactory control is difficult to obtain. McClintock (1915) subjected egg masses of root knot nematode to mercuric chloride, formaldehyde, nicotine sulfate, lime sulfuric acid and concluded that nematodes in egg masses are exceedingly difficult to kill. It has been shown by Taylor (1943) and Stark (1948) that the presence of undecomposed roots containing eggs masses presents one of the most difficult problems in controlling root knot nematode.

The difficulties of killing nematodes in tissue and eggs protected by the gelatinous matrix apply when control of root knot nematodes is attempted around the roots of living plants. The infective larvae in the soil are readily killed by the nematocide. In most cases the soil populations build up to a high level within a short time following treatment because of the numbers of nematodes escaping the action of the nematocide. Satisfactory control under such condition has not been obtained with regularity in experimental plots. The nematocidal chemicals that are presently available for use around the roots of living plants are not satisfactory in all respects. In order to control nematodes adequately by treatment of living perennial plants, it would be desirable to have a nonphytotoxic systemic nematocide or a nonphytotoxic nematocide having long persistence in the soil.

E. Cyst-Forming Nematodes

The nematodes in this belong to the genus *Heterodera*. They differ markedly from the preceding groups in their biology because of the presence of the cyst stage. The infective larvae completely penetrate into host roots, but as the body of the immature female enlarges, it bursts from the root tissue, but remains attached to the feeding site by the head and neck. The female body becomes saclike and completely filled with eggs at maturity. The nematode then gradually changes color from thickening and tanning of the body wall, ceases to feed, and may become detached from the root.

The female body filled with eggs is referred to as the cyst. There are actually two types of biologies within this genus. Some *Heterodera* spp. produce a gelatinous matrix into which the female deposits some eggs immediately upon reaching maturity. However, the majority of eggs are retained within the body. This type of biology is characteristic of *H. schachtii* Schmidt, 1871, *H. cruciferae* Franklin, 1945, *H. trifolii* Goffart, 1932, and *H. glycines* Ichinohe, 1952. The eggs that are deposited in the matrix hatch within a short time and the larvae are infective. In all of these species there may be several generations during the season. This is the result, in part, of the hatching of eggs deposited in the gelatinous matrix and of eggs that hatch from the cyst during the same season they are produced.

In some species of *Heterodera* the females do not produce a gelatinous matrix. All of the eggs are retained within the body. These eggs ordinarily do not hatch except in the presence of a hatching factor produced by the roots of certain plants. This means that the major portion of the soil population remains as eggs within the cyst except in the presence of plants producing the hatching factor. *H. rostochiensis* Woll., 1923, the golden nematode of potato, is the best known example of this type of biology.

Allen and Raski (1950) have reported that eggs within the cysts of *H. schachtii* are about as susceptible to D-D mixture as the infective larval stages of root knot nematodes. Mai and Lautz (1953) found the larvae of *H. rostochiensis* to be slightly more susceptible than eggs in cysts. However, satisfactory control of field populations of *Heterodera* spp. are much more difficult to obtain than is the case with *Meloidogyne* spp. Thorne and Jensen (1946) indicate that in Utah and Idaho satisfactory control of *H. schachtii* can be obtained in most instances with 25 gallons of D-D mixture per acre. In other areas, due possibly to other soil factors, satisfactory controls are not regularly obtained. The cyst-forming nematodes present a difficult control problem with nematocides because each cyst represents a discrete colony of nematocide. An important reservoir of infection remains if the nematocide fails to kill the eggs within a single cyst since each cyst may contain several hundred eggs. Most large-scale field treatments are designed to reduce nematode populations to levels that enable the production of satisfactory yields. It appears that the cyst colonies are to some extent responsible for the failure of nematocides to reduce soil populations except when dosage rates are increased, sometimes beyond economic levels.

F. Nematode Distribution in the Soil

One of the main obstacles preventing complete eradication of nematodes is their distribution in the soil. It is now generally recognized that most nematode infestations occur throughout the root zone of the particular crop. To kill all the nematodes present in the soil requires treatment with nematocides that will diffuse or be carried by water to depths of several feet. Root knot populations are greater at depths of 3 to 4 feet than at 1 to 2 feet in many California fields. Under such circumstances commercial control is obtained by killing most of the nematodes in the top 3 feet of soil. However, with perennial crops, e.g., citrus, grapes, and deciduous trees it is frequently necessary to treat with dosage rates that kill nematodes to a depth of at least 8 feet. Kills at such depth can be obtained only by using large amounts of toxicant under proper conditions for the best nematocidal action.

The upper 2 or 3 inches of soil frequently contain living nematodes following fumigation with volatile liquid nematocides such as D-D mixture and ethylene dibromide. This is primarily due to the rapid diffusion of these nematocides into the atmosphere from the surface layer. This rapid loss prevents the attaining of lethal concentrations in the surface layers. The presence of a few nematodes at the surface after treatment is not important with most annual crops. However, they are a problem in nursery crops subject to quarintine regulations and in treatments designed to give the long protection required with the tree and vine crops. Gas-tight covers, water seals, high dosage rates, and split applications can be used if maximum kills are required. Vapam has been reported by Baines et. al. to be very efficient in killing all nematodes to depths of 5 or 6 feet when applied in water drenches.

The presence of compaction layers, plow soles, and large clods can protect nematodes under some circumstances. Thorne (1939) has shown that a plow sole restricts the downward dispersion of calcium cyanide applied for sugar beet nematode control.

VII METHODS OF APPLYING SOIL NEMATOCIDES

The fact that most successful nematocides are volatile liquids with relatively low boiling points, and with few exceptions rather high vapor pressures, has necessitated the development of special types of application equipment. For the treatment of relatively small areas a number of kinds of hand applicators have been devised.

A. Pressure Applicators

Many applicators operate on the principle of a displacement pump and are designed to deliver accurate dosage rates-as well as different dosage rates depending on the length of the stroke. Taylor (1939) developed and was granted a public patent for such an applicator. All subsequent applicators employ the same principle although numerous improvements have been devised. Hand applicators are designed to deliver a measured amount of nematocide into the soil at the injection site. The depth of the injection is governed by the use of probes of varying length. The spacing of the injection sites may be varied depending on soil type, dosage rate, volatility of the nematocide, etc. In using hand applicators it is customary to mark the soil surface so as to properly space the injections. Taylor (1939) and Chitwood (1939) have described methods of determining the most efficient spacings for the injection of nematocides in single injection sites. The method depends upon a knowledge of the effective lateral diffusion of the nematocide in a particular soil type. This is determined by a bioassay method involving nematode kills obtained at varying distances from the injection point. The entire procedure is aimed at securing the most economical dosage rate and spacing for satisfactory nematode control.

In the treatment of large areas the principle of continuous flow is employed rather than injection of the nematocide at a single point. Virtually all of such equipment in use at the present time employs the principle of continuous flow through a tube mounted on a chisel or shank that is drawn through the soil by power equipment. Essentially the same principle is employed when carbon bisulfide is applied to the furrow in front of a plow share. The applicator devised by Neller and Allison (1935) for introducing chloropicrin and carbon bisulfide into the soil has been modified by growers, equipment manufacturers, and commercial companies. However, nearly all presently used applicators are based on the same principle. Neller and Allison (1935) applied constant pressure to the tank containing the fumigant and regulated the flow of the toxicant by the use of spray orifices and the speed of the machine. Recent modifications include multiple shanks, constant pressure supplied by gear pumps operating from a power takeoff, or ground wheel. Dosage rates ae regulated by pressure, size of orifice, and speed of the machine. Such equipment is usually mounted on a tractor capable of pulling as many as twelve or more shanks through the soil. Various refinements have been devised until it is now possible to accurately apply dosage rates as low as 4 to 6 gallons per acre when the chisel points are spaced at 12 inches so that every square foot receives the desired dosage rate.

B. Gravity Flow Applicators

Numerous types of gravity flow applicators have been devised. They are ordinarily employed to treat small acreages since they are relatively trouble free and are obtained at relatively low cost. Gravity flow combined with a constant head in the fumigant container allows for accurate metering of the fumigant. This principle is frequently employed when nematocides are applied with a plow applicator. In this case the nematocide is allowed to flow into the furrow immediately in front of the plow share. Such equipment can be used on both single and multiple bottom plows.

A modification of the gravity flow applicator in which capillary tubing is used in place of the ordinary valve or orifice has made it possible to apply dosage rates as low as 0.5 gallons of liquid nematocide per acre. In this case the rate of flow is governed by the bore of the capillary tube, its length, and the height of the liquid column. The use of this type of applicator depends on a constant speed in order to apply accurate dosages.

C. Solid Carriers

The effectiveness of modern nematocides when impregnated upon solid carriers has been reported by Taylor and Golden (1954) and Sasser and Nusbaum (1954). The latter authors applied ethylene dibromide with vermiculite as a carrier and employed a fertilizer rig for applying the impregnated vermiculite. The principle involved is essentially that used in the conventional liquid applicators, that of applying continuous lines of the nematocide to the required depth in the soil. This method has no advantage for field scale applications other than the possibility of using equipment designed to make fertilizer applications. Field tests conducted in California in 1955, 1956, and 1957 indicate that applications of dibromo-chloropropane impregnated on Attaclay were as effective as similar dosage rates of the liquid fumigant. Greater cost per unit volume of nematocide and the difficulty of obtaining accurate dosage rates compare unfavorably with the liquid applications. Equipment designed especially for distributing desired dosage rates of nematocides on solid carriers is not presently available for large-scale treatments. The packaging of volatile fumigants impregnated on solid carriers also adds to the expense of such treatments. Moreover, the nematocide is rapidly lost if the containers are left open for any length of time.

D. Drenches

The chemical and physical properties of several nematocides preclude their being injected into the soil. These chemicals are characterized by having extremely low vapor pressures and by not diffusing readily through the soil. Some are relatively insoluble in water and must be applied as emulsions. One currently used material, sodium-*N*-methyl dithiocarbamate (Vapam) is very soluble in water and is ordinarily applied as a water solution drenched into the soil. Drenches containing ethylene dibromide, D-D mixture, allyl alcohol, and similar materials in emulsified form have been used under special circumstances to control nematodes in the surface, 2 to 3 inches of soil. The use of nematocides whose movement in the soil is dependent upon water has been generally restricted to special uses such as preplanting treatments for trees, vines, and nursery beds. Their use may depend on the presence of pathogenic organisms such as fungi, although nematodes may also be present and controlled by the treatment.

E. Over-all and Strip Treatments

Soil treatments may be applied over the whole of the area or in narrow strips where the row of crops will be. Area, over-all, or broadcast treatments involve treating the entire area to be planted. The nematocide is usually applied in continuous lines at the desired depth and spacing so as to effectively treat as much of the soil as possible. Such applications may also be made by hand injections. Conventional treatments are usually made by applying the volatile liquid nematocides to a depth of 8 inches with 12 inch spacing of the chisel shanks. For special problems the depth of application and the spacing may be varied. Baines et. al. (1957) recommend a spacing of 18 to 20 inches and a depth of 10 to 14 inches for preplanting treatment of old citrus soil. If, however, the soil contains large amounts of clay these authors recommend a spacing of 12 to 15 inches. These treatments involve the application of 50 to 240 gallons of D-D mixture per acre, the amount used depending upon soil type and the depth of the soil.

Row or stip treatments are designed to reduce or eliminate nematode populations in the immediate planting area. Row treatments have been developed over a long period of time and for a variety of situations. The treating of planting sites with carbon bisulfied for the control of *Armillaria mellea* embodies the same principles. Among the first and most widely used of the row or strip treatments is that used in pine-apple fields in Hawaii. Chloropicrin was used in this manner prior to the discovery of D-D mixture. The procedure in this case involved the injection of chloropicrin into the immediate soil area upon which plants were to be grown. The planting area was covered subsequent to injection of the chemical by specially treated paper to aid in retaining the fumigant in the soil. The area between beds was left untreated. D-D mixture has been used in the same manner since it became available. The basic idea is to provide protection for young plants for a period of time that will allow them to become well established before they are subjected to attack by large populations of nematodes.

The protection afforded by row or bed treatments is not equal to that obtained by area treatments, but row or bed treatments are frequently more economical in terms of net profit. It is common practice to employ row or bed treatments to control nematodes on cotton, tomato, bean, melons, and similar crops. Dosage rates are dependent upon row or bed spacing, soil type, the nematocide used, and the susceptibility of the particular crop to nematode damage. Generally row treatments are not employed when the marketable portion of the crop is subject to nematode attack, e.g., carrots, Irish potato. Row treatments are not employed to treat land that is to be used to grow perennial crops.

F. Special Procedures

There has been limited use of nematocides to control *Ditylenchus* spp. Adults and larvae are usually found in the upper layer of the soil. This is the area where most volatile nematocides are least effective and high dosage rates in addition to special procedures are required to secure adequate control which involves covering the soil with gas-impervious covers or using split treatments. The latter procedure is carried out by applying one-half the dosage rate and then plowing after a 2 or 3 day interval and injecting the other half. Dallimore (1955) has reported that a scraper attachment mounted in front of the plow provides a successful means of adequately treating the upper 3 or 4 inches of the soil. The nematocide is injected and after an interval of 4 to 5 days the soil is plowed, at which time the scraper pushes the top soil into the furrow. The plow then covers the soil in the furrow where it is exposed to the action of the nematocide. This method is expressly for the purpose of treating the upper layers of the soil and seems to be effective only against such nematodes as *Ditylenchous dipsaci* and *D. destructor* that are found in-upper soil layers. The method has no appreciable advantage for use against species that are distributed to considerable depths in the soil, e.g., root knot nematodes, cyst nematodes, burrowing nematode, and root lesion nematodes.

VIII. METHODS OF CONFINING SOIL FUMIGANTS

The effectiveness of volatile nematocides as a means of controlling nematodes in the soil is to a large extent dependent upon the retention of toxic concentrations of the vapors in the soil. Certain of the physical properties of the soil, e.g. porosity, clay content, water content, compaction, and organic content, influence retention of the vapors. If gas-impervious covers or water seals are not used to insure the retention of toxic concentrations of the toxicant, then the actual manipulations of the soil and its condition at and subsequent to treatment are extremely important. Various devices have been used to smooth the surface of the soil following the injection of nematocides with chisel-type applicators. Harrowing, dragging a chain, heavy iron bar, or wooden drag behind the applicator tends to seal the soil surface resulting in better retention of toxic vapors. In most situations, more effective retention seems to result if a ring-roller or similar implement is pulled immediately after the applicator. In addition to smoothing the surface, the ring-roller has the added advantage of compacting the surface layer. Consistently good results have been obtained in California using this method on experimental plots.

Water seals and gas-impervious covers have been mentioned previously and their main effect is to prevent rapid loss of the nematocide from the soil. Polyethylene covers are regularly employed when soil is treated with methyl bromide. They may also be used with chloropicrin when a high degree of control is desirable. Similarly, water seals increase the effectiveness of chloropicrin. It has been observed in the field that rain immediately following injection of such materials as D-D mixture and ethylene dibromide results in longer retention of the gases with attendant increased control of nematodes. Phytotoxicity to subsequent plantings has also been observed when the interval between treating and planting was not increased to allow more time for the diffusion of the toxicant from the soil.

IX. TREATMENT OF LIVING PLANTS (POSTPLANTING)

The injection of volatile liquids into soil around the roots of living plants has been attempted with a variety of materials. D-D mixture and ethylene dibromide are not satisfactory because they are phytotoxic. Of the presently used volatile liquid nematocides only 1, 2,-dibromo-3-chloropropane seems to possess low enough phytotoxicity to enable its use around the roots of living plants at dosage rates that are nematocidal. Lownsbery and Sher (1958) have reported significant increase in growth of young walnut trees side-dressed with this material when the trees were planted in soil previously treated with a nematocide. They indicate that growth rates were not satisfactory for trees growing on preplanting fumigated soils, but that subsequent side-dressing treatments increased growth and reduced nematode populations. The use of this material for the purpose of side-dressing established plantings is still in the experimental stage. It is probable that some of the unsatisfactory results that have been obtained are associated with attempts to rejuvenate older plants that had extremely poor root systems.

Satisfactory results have been reported from some experiments using dibromochloropropane to side-dress such crops as cotton and beans for the control of root knot nematode. However, about an equal number of failures have also been encountered in the field. In situations where increased yields have resulted, the increases do not compare favorably with those obtained with preplanting treatments with equal amounts of the same material. In some instances cotton has been severely injured by side dressing treatments also resulted in lower yields than were obtained from untreated areas.

The organic phosphate, *o*-2, 4-dichlorophenyl-*o*, *o*-diethylphosphorothioate, (V-C 13) has been reported to be a satisfactory chemical for treatment of living plants. It is used primarily as an emulsion and drenched into the soil. It is reported to be chemically stable in the soil, has a very low vapor pressure, and its movement in the soil depends on the movement of the water emulsion. Christie and Perry (1958) have discussed various aspects of its use and indicate that it is an extremely promising material for use around living plants. V-C 13 applied as a drench is reported to improve the growth of turf by controlling associated pathogenic nematodes.

Tarjan (1955) has reported that aqueous emulsions of 3-*p*-chlorophenyl-5-methyl rhodanine completely control root knot nematode infections when applied to potted tomato plants. He reported that the chemical reduces nematode populations in the soil around the roots of strawberry plants when it is mixed in the dry form with the soil.

X. PHYSICAL FACTORS INFLUENCING THE EFFICIENCY OF NEMATOCIDES

The soil is the greatest barrier to the successful use of nematocides. Over a period of years, a large body of evidence accumulated indicates that the success of conventional fumigation treatments is largely governed by the soil to which they are applied. It is also recognized that little can be done to change those characteristics of soils that are unfavorable to the effective use of nematocides. The amount of clay and/or the amount of organic matter are the limiting factors that are fairly well fixed and the difficulties inherent in their presence respond only slightly to any practice that can be reasonably carried out in the field.

It is unfortunate from the practical aspects of killing nematodes, under field conditions, that most of the precise studies on the characteristics of soils that inhibit the diffusion of gases have been made on dry soil. However, there can be no doubt that some of the soil factors discussed in the following sections are operative, although in several instances experimental evidence is lacking or incomplete concerning the mechanisms that influence or restrict diffusion.

A. Sandy Soils

Sandy soils and sandy loam soils have relatively low organic and clay contents and are representative of soils in which volatile nematodes as well as other types of nematocides are most effective in controlling nematodes. The effectiveness of volatile nematocides in these soils is directly attributable to the relatively unrestricted diffusion of the toxic vapors. It is known from the work of Hagan (1941), Call (1957), Wade (1954, 1955), and Stark (1948) that carbon bisulfide, ethylene dibromide, and chloropicrin are highly sorbed when they are in contact with dry soil. Hagan (1941) has indicated that soil moisture is the most important factor limiting diffusion. Call (1957) and Wade (1955) have shown that ethylene dibromide sorbed by dry soil can be released by the addition of moisture. It therefore, follows that moisture must be present in order to prevent the excessive sorption that takes place in dry soil. Lear (1956) has shown that Vapam applied as a concentrate solution to surface soil is retained in the surface of wet soil and is carried out of the surface of dry soil by addition of water. Most consistent result are obtained by applying the complete treatment as a drench.

In the field, volatile nematocides are most soils. It has been observed that these nematocides are most effective when they are applied to sandy loam soils that are at or near their moisture- holding capacity. We must, therefore, assume that effective treatment of such soils requires a balance between water content and pore space that enables the gaseous phase of the nematocide to diffuse and reach toxic concentrations throughout the soil. It is further assumed that the toxic vapors go into solution in the water phase and nematodes are killed by contact with water solutions.

B. Clay Soils and Organic Soils

There are no precise data to indicate the exact clay or organic content of the soil that will restrict diffusion to the point where effective control is not obtained. If we assume as Call (1954) and Wade (1955) have indicated for ethylene dibromide that water inhibits adsorption of this material on soil particles, we must look for other causes for poor diffusion of the gaseous phase of volatile nematocides. As indicated in Table 9.1, the volatile nematocides are slightly soluble in water, varying from 0.09 gam. per 100 ml. of water for methyl bromide to 0.43 gm. per 100 ml. of water of ethylene dibromide. If a conventional dosage rate of 20 gallons per acre of D-D mixture is applied as a broadcast or area treatment, each cubic foot of soil receives a dosage rate of 1.87 ml. of the nematocide. Under field conditions there is sufficient water in each cubic foot of soil to dissolve the entire amount of D-D mixture (sol. 0.27 gm. per 100 ml. of water) applied. It seems likely that in clay and organic soils, having field capacities exceeding 25%, large amounts of the gaseous phase of the fumigant are dissolved in the water surrounding soil particles at the point of injection, thus restricting the degree and speed of diffusion. In addition, the presence of smaller pore spaces in clay soils and the fact that adsorbed water further decreases the size of the pores leads one to the conclusion that clay soils are not successfully treated with volatile nematocides because of the restrictions on movement imposed by water present in these soils. It is highly probable that other factors are operative in the different types of clays and in organic soils where the situation is complication by factors that are not well understood at present. Goring and Youngson (1957) have shown that diffusion of ethylene dibromide is restricted by the addition of organic material to sandy soil and that different materials have different effects.

In treating clay and organic soils, poor or limited diffusion can be overcome to some extent by the application of higher dosage rates. However, there are economic limitations that sometimes make it impractical to apply dosage rates that the are effective. This is clearly the case with sugar beet nematode in clay soils as shown by Allen and Raski (1950). Similarly, the change in dosage rate that is required for various types of soil is indicated by the recommendation of Baines et. al. (1957) of 60 gallons per acre of D-D mixture as a preplanting treatment to control citrus nematode on sandy soil and 250 gallons per acre to achieve comparable kills in clay soils.

C. Moisture

As indicated in the preceding section, it is impossible to discuss soil type without reference to moisture unless dry soils are being considered. However, the practical use of volatile nematocides never involves completely dry soils. We must, therefore, consider the conditions of moisture that are generally thought to be most advantageous for the use of volatile nematocides. It is obviously not possible to have precise data for every soil type encountered in the field. In sandy and sandy loam soils (moisture equivalent 5 to 12%), best results are obtained when treatments are made at or near field capacity. Soils with high clay content cannot be cultivated when at field capacity and it is necessary that these soils be treated at a time when they are below their field capacity. It is recommended that these soils should be treated when they are in suitable condition for cultural operations and in seed-bed condition so far as tilth is concerned. Van den Brande et. al. (1954) have indicated that adequate soil moisture must be present in order to effectively reduce *Heterodera rostochinsis* populations. Practically no data are available to make possible a comparison between dry and wet soils. There can be no doubt that reasonable soil moisture is desirable but it is not known whether soils with high clay content could be treated successfully if moisture content is near the wilting point. It is likely that the phenomenon of adsorption by soil particles becomes operative if the moisture content is reduced below this level.

D. Temperature

The proper temperature for the application of nematocides is a controversial subject probably because no single investigation has been undertaken that includes factors that are operative under field conditions. McClellan et. al. (1949) have shown that such materials as chloropicrin,

D-D mixture, and ethylene dibromide are more effective at 85 to 95° F. than at lower temperature. However, these experiments do not have application to field situations since the containers prevented escape of the nematocide. furthermore, the length of the exposure period was very short so that at lower temperature not only would one expect slower diffusion, but also the activity of the nematodes would be reduced. Although laboratory data may indicate greater nematocidal activity at higher temperatures, there is also more, rapid diffusion from the soil into the atmosphere.

Successful nematode control has been obtained by the application of some nematocides to field soils at temperatures as low as 37° F. at the 8 inch depth. It appears that if all other conditions are favorable, there is considerable latitude in the temperature requirements of nematocides. This is a fortunate circumstance since treatments must sometimes of necessity be applied in spring and fall at temperatures that are sometimes lower than those recommended on the basis of laboratory experiments. Van den Brande et. al. (1954) have shown that treatments made at 36.5° F. den are as effective as those made at 59° F.

Although diffusion may be restricted at low temperatures, escape from the soil is likewise restricted so that satisfactory control is obtained, but the period of time to achieve such control is necessarily longer Goring and Youngson (1957) have shown that ethylene dibromide does not diffuse in soil at 50° F. but they indicate that the nematocide remains in the soil and kills as the temperature rises.

E. Preparation of the Soil

Aside from adjusting soil moisture to the optimum level and waiting for favorable temperatures, it is possible to improve the conditions of the soil by cultural operations. Diffusion and resultant kill of nematodes is enhanced when the soil is in a friable condition. Absence of large clods, compaction, and rough surface tend to provide conditions conducive to obtaining maximum efficiency of volatile soil fumigants. Undecomposed crop residues regardless of whether or not they contain nematodes or nematode eggs are detrimental. These resideus tend to collect upon the chisels and to make the surface of the soil rough, which in turn exposes more soil surface and increases diffusion into the atmosphere. Stems or roots that protrude from the surface of the soil act as chimnesy for the escape of the gaseous phase of the fumigant. Excellent results with the volatile soil fumigants are usually obtained when the soil is properly prepared and is free of crop resideus, large clods, the surface smooth, and the moisture content is at a favorable level.

F. Posttreatment Conditions

Atmospheric conditions at, during, and following the application of volatile soil fumigants are very important in determining the ultimate effectiveness of a nematocidal treatment. Immediately following treatment the soil surface should be smoothed to fill any channels that may be left by the chisels as they are pulled through the soild. It is preferable to pull a ring-roller behind the fumigant applicator. Although wind and temperatures following treatment cannot be controlled, it is important that the soil mass should remain undisturbed for a period of time following treatment the soil surface should be smoothed to fill any channels that may be left by the chisels as they are pulied through the soil. It is preferable to pull a ring-roller behind the fumigant applicator. Although wind and temepratures following treatment cannot be controlled, it is important that the soil mass should remain undisturbed for a period of time following fumigation. Any disturbance of the soil surface is apt to increase diffusion from the soil.

If rain immediately follows treatment it will tend to seal the soil surface and increase the efficiency of the treatment. If highly phytotoxic materials such as D-D mixture and similar materials are used, and rain follows the treatment, there is a possibility that retention of the toxicant may result in phytotoxicity if planting follows at the recommended interval. Under such conditions the interval between treatment and planting should be extended. This is extremely important where recommended intervals are based on minimum times under optimum conditions.

XI. THE MODE OF ACTION OF NEMATOCIDES

Practically no information is available concerning the mode of action of any nematocide. Nearly all of these chemicals are biocides and do not show enough specificity to indicate a mode of action that would be peculiar to nematodes. It is assumed that since the phosphate insecticides are toxic to nematodes, these materials act upon the nervous system and the enzyme cholinesterase. It has been suggested by Chitwood and Chitwood (1950) that effective nematocides must be materials that change the permeability of the cuticle and should also be soluble in or dissolved by fat solvent. However, there are little or no data available to support the necessity of penetration through the cuticle. It is easily observed when nematodes are exposed to water solution of vital stains that penetration is most rapid through the oral opening, phasmids, vulva, and anus. There is no reason to assume that most nematocidal chemicals do not enter the body of the nematode through these avenues as well as through the cuticle. It is reasonable to assume that chemicals that are ovicidal must penetrate through the egg shell in order to be effective. Ethylene dibromide which is not as ovicidal as D-D mixture or chloropicrin does not kill as rapidly as the latter materials. This might indicate that penetration through the cuticle or the egg shell is not as rapid with ethylene dibromide as with the other materials. Data are not available concerning the ovicidal properties of 1, 2,-dibromo-3-chloropropane but since it does not kill larvae of root knot nematode as rapidly as ethylene dibromide it can be assumed that this nematocide might also have rather poor ovicidal properties.

The gelatinous matrix produced by root knot nematodes, citrus nematode, reniform nematode, and some cyst nematodes protects eggs from nematocidal action. This has been shown when treatments of living roots are attempted. It has been mentioned previously that the cyst wall does not afford an effective barrier to some nematocides and failure to control *Heterodera* spp. is probably not due to this factor.